The genus Inga

Utilization

T0141433

FORESTRY RESEARCH PROGRAMME

CGIAR

**CONSULTATIVE GROUP ON INTERNATIONAL
AGRICULTURAL RESEARCH**

INTERNATIONAL CENTRE FOR RESEARCH IN AGROFORESTRY
supported by the
Consultative Group on International Agricultural Research

The genus *Inga*
Utilization

T.D. Pennington
&
E.C.M. Fernandes
(Editors)

The Royal Botanic Gardens, Kew
1998

First published 1998

ISBN 1 900347 58 X

Production Editor: Ruth Linklater; cover design by Jeff Eden, page make-up by Margaret Newman, Media Resources, Information Services Department, Royal Botanic Gardens, Kew

Cover photograph: Sue Cunningham

Printed in The European Union
by
Continental Printing, Belgium.

CONTENTS

LIST OF CONTRIBUTORS

Ilse L. Ackerman,
Department of Soil, Crop, and Atmospheric Sciences (SCAS),
Cornell University,
Ithaca,
NY 14853, USA.

Julio C. Alegre,
International Centre for Research in Agroforestry,
INIA-PNIACT,
Carretera Federico Basadre KM. 4.200,
Pucallpa, Peru.
e-mail: j.alegre@cgnet.com

Dale E. Bandy,
International Centre for Research in Agroforestry,
INIA-PNIACT,
Carretera Federico Basadre KM. 4.200,
Pucallpa, Peru.
e-mail: d.bandy@cgnet.com

Erick C.M. Fernandes,
Tropical Cropping Systems & Agroforestry (SCAS),
Cornell University,
Ithaca,
NY 14853, USA.
e-mail: ecf3@cornell.edu

Michael R. Hands,
Department of Geography,
University of Cambridge,
Downing Place,
Cambridge CB2 3EN, UK.

Jorge León,
Apartado 480,
San Pedro,
Montes de Oca,
Costa Rica.

Ellen L. McCallie,
Department of Soil, Crop, and Atmospheric Sciences (SCAS),
Cornell University,
Ithaca,
NY 14853, USA.

Richard J. Murphy,
Department of Biology,
Imperial College of Science, Technology and Medicine,
Prince Consort Road,
London SW7 2BB, UK.
e-mail: r.murphy@ic.ac.uk

David A. Neill,
Missouri Botanical Garden,
P.O. Box 299,
St. Louis,
Missouri 63166, USA.
e-mail: neill@mobot.org

Giovanni Onore,
Department of Entomology,
Pontificia Universidad Católica del Ecuador,
Apartado 17-01-2184,
Quito,
Ecuador.
e-mail: gonore@puceu10.puce.edu.ec

Terence D. Pennington,
Royal Botanic Gardens,
Kew, Richmond,
Surrey TW9 3AB, UK.
e-mail: t.pennington@rbgkew.org.uk

Nixon Revelo,
Fundacion Jatun Sacha,
Casilla 17-12-867,
Quito,
Ecuador.
e-mail: dneill@jsacha.ecuanex.net.ec

Richard K. Robinson,
Department of Food Science & Technology,
The University of Reading,
P.O. Box 226,
Whiteknights,
Reading RG6 6AP, UK.
e-mail: r.k.robinson@afnovell.reading.ac.uk

Giovanna Rodriguez,
Department of Entomology,
Pontificia Universidad Católica del Ecuador,
Apartado 17-01-2184,
Quito,
Ecuador.

John C. Weber,
International Centre for Research in Agroforestry,
INIA-PNIACT,
Carretera Federico Basadre KM. 4.200,
Pucallpa, Peru.
e-mail: j.weber@cgnet.com

P.Y. Yau,
Department of Biology,
Imperial College of Science, Technology and Medicine,
Prince Consort Road,
London SW7 2BB, UK.

INTRODUCTION

Inga has a history of utilization extending back 2000 years, when it was cultivated for its edible fruit by the pre-Colombian inhabitants of Peru and, more recently, as a shade tree for coffee, cacao and tea. These aspects are discussed in Chapters 1 and 7 of this volume. *Inga* species are still widely grown to-day especially in the Andean countries and in Central America. Each region has its preferred species of edible *Inga* and in season they are sold in large quantities in the markets. Some species, such as *I. edulis* and *I. feuillei*, have been distributed by man and subjected to selection for fruit and seed size and they are now found far from their native area.

Inga is also widely used today as fuelwood because of its rapid growth, ability to withstand repeated coppicing and it burns well without producing too much smoke. As a high proportion of rural people still use fuelwood for cooking, *Inga* is becoming increasingly important in areas where the native forest has been cut down.

The contributions in this book, while also covering traditional uses of *Inga*, are focused on the restoration of degraded acid soils, and its utilization in agroforestry and forestry systems.

Large areas of land in lowland and montane tropical America have been deforested, cultivated or put down to pasture for a few years and then abandoned because of poor fertility. Agroforestry systems provide a means of bringing back such land into productive use, thereby reducing pressure on the remaining forest. In recent years attention has focused on *Inga* as a potentially useful multipurpose plant in such agroforestry systems (see Chapters 4 and 8). With forest destruction carried out mainly by farmers seeking land to cultivate subsistence crops, emphasis is being placed on systems which will not only take some of the pressure off the cutting of virgin forest but also enable land to be cultivated for longer than is the case with the usual slash and burn. Alley cropping systems are discussed in Chapter 5, and many *Inga* species seem to be ideally suited as multipurpose components of such systems particularly on poorly drained acidic soils where other leguminous agroforestry trees, such as *Calliandra* and *Leucaena* do not perform well.

The characteristics of *Inga* which make it so promising are summarized below.

1. **High species diversity and great ecological range**. *Inga* is a huge genus of around 300 species widely distributed and common throughout lowland and montane regions of humid tropical America. Some species (e.g. *I. marginata*) have an altitudinal range of over 2000 m, while different provenances of other species, such as *I. vera*, can tolerate an everwet climate with 5000 mm of annual rainfall or a strongly seasonal climate with a 5–6 month dry season and greatly reduced rainfall. *Inga* is an ubiquitous group and each locality has its own set of species adapted to the local conditions. Local species can usually be found for local uses thereby avoiding dependence on a single species with the associated problems of pests and diseases.

2. **Rapid growth**. Many *Inga* species are fast growing light demanding plants which have the ability to compete successfully with weedy secondary vegetation. The growth characteristics of *Inga* species are discussed in Chapter 2.

3. **Rapid Germination**. *Inga* is easily grown from seed, with normal germination rates of 95–100%. Given moisture and shade they germinate within a week or two of planting. The requirements for seed handling, storage and germination are described in Chapter 11.

4. **Tolerance of poor soils**. Many *Inga* species are well adapted to infertile, red acidic soils, such as are found over wide areas of the lowland humid tropics of Amazonia. Such species as *I. edulis* and *I. marginata* flourish under very low pH conditions, which other legumes cannot tolerate. Other *Inga* species do well on poorly drained or periodically flooded sites. They are particularly useful for rehabilitating compacted pasture land (Chapter 9).

5. **Improving soil fertility through nitrogen fixation and mycorrhizal activity**. All species of *Inga* so far investigated produce root nodules containing nitrogen fixing bacteria (Chapter 4). Crops grown in combination with *Inga* benefit from the release of nitrogen and also from a sustained release of nutrients from the slowly decomposing leaf mulch. The permanent mulch beneath *Inga* trees has the effect of causing rooting to be raised up into the surface layers (similar to the situation found in natural forest) above the region of aluminium toxicity. *Inga* roots form associations with mycorrhizal fungi, which probably provide the means by which *Inga* plants are able to recycle phosphorus which is unavailable to non-mycorrhizal plant species on tropical acid soils. (See Chapters 4 and 5). The permanent leaf mulch reduces the soil surface temperature to the levels found in natural forest so enabling the germination of crop seeds sown in it.

6. **Good shade trees**. All *Inga* species have essentially the same branching pattern which, when growing in an open situation, gives rise to the characteristic broad umbrella-shaped crown. This shape makes *Inga* an excellent shade tree for such crops as coffee, cacao and tea, which require a partial shade with sunflecks. The amount of shade cast varies from one species to another so different *Inga* species are preferred for different crops (Chapter 7).

7. **Weed control**. The shading effect of *Inga* can be used to good effect for weed control and recuperation of abandoned pasture or secondary vegetation. Some species, such as *I. oerstediana*, combine very fast growth and competitive ability with very large leaves. The leaves which fall throughout the year are relatively slow to decompose and soon form a long-lasting mulch below the tree and this, combined with the shading effect of the *Inga* crown, soon depresses the growth of all vegetation below the trees and within a year or two eliminates it, producing a clean forest soil which can be brought into productive use. The effectiveness of *Inga* for weed control is discussed in Chapters 2 and 5.

8. **Coppicing ability**. All *Inga* species investigated have the ability to withstand coppicing from an early age. In Chapters 2 and 5 the results of biomass production under a coppice regime are described. All species were found to coppice well and the high quantities of biomass produced by coppicing regularly can be used as green manure and for weed control on crops grown in the rows between the *Inga* trees (Chapter 5).

9. **Fuelwood**. Throughout Central America and western South America, where a large proportion of the population still rely on wood for cooking, *Inga* species are usually cited as a preferred fuelwood. The reasons for this are several: its fast growth, tolerance of coppicing and a wood than burns well without producing a lot of smoke. Some species, such as *I. punctata*, are particularly good, producing a lot of fuelwood-size branches (3–4 cm diameter and more) within 6 months of coppicing and having wood that is quite dense. Although no information is available at present it is likely that some *Inga* species would be good charcoal producers. The characteristics of *Inga* wood for fuel are discussed in Chapter 3.

10. **Biological Interactions**. All *Inga* species have small nectar-producing glands on the leaves. These attract a wide range of insects to the plant, especially ants. The direct effect of these visiting insects is that they protect the *Inga* plant against herbivores. However, there is also an indirect benefit in that the visiting insects may also parasitize pests living on crop species grown among the *Inga* trees. In this way *Inga* has been successfully used as a nurse crop for other timber species, such as *Swietenia macrophylla,* which is normally heavily parasitized by a shoot borer *Hypsipyla* (Chapters 8 & 9). The association of *Inga* with a wide range of insects can result in both positive and negative impacts on associated crops. There is a good potential for enhancing integrated pest management (IPM) strategies in systems involving *Inga* (See Chapter 8).

11. **Edible fruit**. All species of *Inga* have edible fruit and many are protected and cultivated for this reason (See Chapters 1 and 10). At certain times of the year the fruit form an important item of commerce in the local markets so their cultivation can provide a useful source of additional income for farmers and peasants. Each country in Central America and the Andean region has a unique group of species with edible fruit extending from the lowlands up to 3000 m altitude.

12. **Timber**. Although little information is available on the utilization of *Inga* as a timber producer it appears that some species, such as *I. alba* and *I. leiocalycina* which grow to a large size, are used locally as a source of sawn timber. Species such as these, which have straight cylindrical bole and reach a metre or more in diameter, should be brought under trial to find out more about their growth rates and other characteristics.

CHAPTER 1. HISTORY OF THE UTILIZATION OF *INGA* AS FRUIT TREES IN MESOAMERICA AND PERU

JORGE LEÓN

Inga trees are used for their fruits throughout their geographical range from Mexico to Uruguay. The useful part is the pulp (sarcotesta) that surrounds the seeds. It is a watery, soft, slightly sweet, generally white tissue, which in *I. feuillei* contains 84% water, 13% carbohydrate, 0.7% fibre, 0.6% protein (Collazos *et al.*, 1957, see also Pennington & Robinson, this volume).

Within the area of distribution of *Inga*, there are some regional concentrations of useful species. One is in the Amazon, where some species have been improved by human selection (Ducke, 1946). Among them are two outstanding species that deserve to be introduced and tried in other tropical regions: *I. cinnamomea* and *I. edulis* (Cavalcante, 1988). There are other promising genotypes, like a variety of *I. fagifolia* (= *I. laurina*) known only in cultivation, and *I. macrophylla*. A minor centre in the utilizaton of *Inga* as fruit trees is the southern part of Brazil, where *I. affinis* (= *I. vera*), *I. barbata*, *I. sessilis* and *I. uraguensis* (= *I. vera*) are planted or collected (Hoehne, 1948; Rodriguez Mattos, 1978).

Another centre is Mesoamerica: central Mexico to NW Costa Rica. In the registers of plants cultivated in Mesoamerica before the arrival of the Europeans, there is only a dubious reference to *Inga*. It is the description of Dr Francisco Hernandez who lived in Mexico from 1570 to 1577, of a tree in Michoacan, locally called chala, "with pods four fingers wide and a palm and a half long...with fifteen oblique seeds as wide as the pod, covered with a white membrane, villous as silk or like the hairy cover on unripe almonds. The seeds are green and similar to broad beans. This fruit is eaten cooked; although it is almost tasteless and odourless, it has a certain sweetness" (Hernandez, 1959). It may be possible that Dr Hernandez is referring to *I. jinicuil*, which grows in Michoacan. In its utilization, he may confuse the almost tasteless cooked seeds and the sweet tasting pulp that covers them. It is probable that this species has been selected and cultivated by the Indians for a long time. The specific name derives from a nahuatl word, and means 'twisted feet', for the shape of the fruits (Cabrera, 1978). In the lists of plants cultivated in ancient Mexico (Dressler, 1953; Luna Cavazos, 1990), *I. jinicuil* does not appear.

In Guatemala, El Salvador and Costa Rica *I. jinicuil* (formerly regarded as a distinct species *I. paterno*) is used to shade coffee and is also planted for its fruits. The fruit is among the best in the genus, the seeds are up to 6 cm long, with thick, translucid and sweet pulp. The roasted seeds of *I. jinicuil*, according to Roskoski (U.S. National Research Council, 1989) are sold as peanuts outside theatres to moviegoers. In El Salvador, the seeds are cooked, bleached, salted and eaten as a salad (Standley & Steyermark, 1946); in Costa Rica, they are cooked, cut into little cubes and eaten as "picadillo". No references have been found on nutritive value of these seeds.

The *PACAE* (*INGA FEUILLEI*) IN PERU

Inga feuillei grows on the Pacific slopes of South America from Colombia to Chile, only in cultivation (Pennington, 1997). On the coast of Peru, the *pacae* has been cultivated for more than 22 centuries, and under human selection, it has become the most advanced fruit tree in the genus.

The coast of Peru is one of the most arid regions of the world, mainly due to the effect of the maritime currents that run south to north, and cool the surface of the ocean forming dense fogs but no rain. The rainfall is practically nil, and occurs in erratic fashion in the northern section, especially as a result of the climatic condition called "El Niño". The El Niño phenomonen which occurs every 2 to 5 years, can bring heavy rains to the normally arid coast of Peru and Ecuador. The desert belt is crossed, from the boundary with Ecuador to Chile, by some fifty rivers that descend from the Andes, first through narrow gorges, then expanding in alluvial deposits, some of considerable extension. Each of these valleys is an oasis, separated from the others by desert areas in which the only vegetation are patches of *Tillandsia* and cactus.

The coast was first settled by people who lived off the animal and plant products that the ocean supplies in enormous quantities. There is no proof that these people developed agriculture in the valleys, or if these were occupied by settlers that came from the Andes or the Amazon, in the latter case, through the lower mountains of northern Peru. But one factor in the development of agriculture on the coast seems to be clear, that it was not based on local domestications, as has been proposed by various archaeologists (Lanning, 1967; Lumbreras, 1974). The local vegetation, found only along the rivers, is very poor in species, as it is along the whole coast (Weberbauer, 1945). None of them could be considered as the ancestor of the crops cultivated in the valleys. Chilli peppers for instance, came from Bolivia or the eastern part of the Andes, corn came from Mesoamerica; from the Amazon, cassava, pineapples, peanuts, annatto (*Bixa orellana*), palillo (*Campomanesia lineatifolia*); from the Andes, chirimoya (*Annona cherimola*), achira (*Canna edulis*), jícama (*Pachyrhizus ahipa*), beans (*Phaseolus* spp.), lucuma (*Pouteria lucuma*), possibly *pacae* (*Inga feuillei*); from northern South America (Colombia–Ecuador) *Cucurbita* spp., ciruela del fraile (*Bunchosia armeniaca*), guava (*Psidium guajava*), pepino (*Solanum muricatum*), cotton (*Gossypium barbadense*), guanabana (*Annona muricata*), avocado (*Persea americana*); from Argentina, *Cucurbita maxima* and from Chile, *Aristotelia maqui*. Apparently only *Begonia geraniifolia* and *Canavalia plagiosperma* could be considered as coastal domestications.

The development of agriculture on the coast was possible only through irrigation. Nowhere else on the continent were the irrigation systems so extensive and efficient as in coastal Peru. As in Mesopotamia, the irrigation, crop introductions and management of the produce, were the physical base for establishing and maintaining a permanent agriculture, no matter what political or ethnic group was in power.

ARCHAEOLOGICAL REMAINS

The climatic and soil conditions of the coastal area of Peru are probably the most favourable for the preservation of plant remains, especially branches, leaves and seeds. In *pacae*, all these organs as well as pods are remarkably well preserved. Leaves are especially abundant in graves, as they were used as filling in the mummy bundles. In the first studies of *Inga* leaves (Rochebrune, 1879; Wittmack 1888), the possibility arose that the leaves might belong to more than one species. *Inga feuillei* was accepted as one, and it was suggested that the other could be *I. fasciculata* (Rochebrune, 1879); this name is now a synonym of *I. oerstediana*, a species which in Peru grows on the eastern slopes of the Andes and the Amazon, with short, almost inedible, cylindrical fruits. Harms (1922) also separates two entities, based on leaf characters. The first, which he assumes to be *I. feuillei*, has broader leaflets; in the other, there are 3–4 pairs of leaflets on a winged rachis close to 13 cm long. The leaflets are more lanceolate and narrow, 4–12 cm long, 1–3 cm wide, and less pubescent than in the first entity. At present, two variants of *I. feuillei* are known in central Peru, one with hairy leaves, the other more or less glabrous (Pennington, 1997).

Pods or fragments of valves are very common in tombs, placed alone or in groups at the base of the mummies. In Ancon, one pod was found lying across the chest of a five year old boy (Wittmack, 1888). Also in Ancon, in several underground tombs, the rush mats forming a kind of roof, were made of trunks or limbs of *pacae*, a very durable wood (Safford, 1917).

The chronological table for the coast of Peru (Willey, 1971) starts at 6000 BC with the Preceramic Periods that end at 1800 BC. Within these periods, remains of *pacae* appear at Los Gavilanes site, between 2700–2200 BC and at Ancon, around 1750 BC. In the next sequence, called Initial Period-Early Horizon, that ends at 200 BC, there are remains from Gramalote, 1590–110 BC; Garagay, 1400–600 BC; Cardal, 1180–850 BC; Pampa Rosario, 810–450 BC and San Diego, 550–295 BC, (data from Pearsall, 1972). There are also reports from Pachacamac, Nazca, Ica and Paracas (Towle, 1961).

CERAMIC REPRESENTATIONS

The most complete representation of prehistoric crop plants is found in Peruvian ceramics, textiles, stone and metal artefacts. The best of them, from an artistic point of view, are in the Mochica pottery (100–800 AD), but the most common and the most interesting for their botanical information, are in the Chimu pottery (1100–1470 AD). Although in the Chimu ceramics "there is a drab sameness, a subtle lack of fine proportions and a carelessness in the execution" (Willey, 1971), they are an invaluable source in our knowledge of prehistoric crops. In the Chimu there are ceramics representing most of the crops that were planted in the coastal area in their different varieties, including rare mutations, like the bicoloured gourds.

Pacae pods are represented in several forms:

a) as copies of the fruit of natural size (Fig. 1A), probably using the pod as a matrix, covering it with plastic clay to obtain a two piece mould. The pieces of the mould are separated along the narrow sides of the pod and when copies are made, the marks left by the joint have to be finished by hand. This produces some distortions in the surface of the narrow sides of the pod. There are also representations of pods in typical twisted positions (Fig. 1B).

b) flat stirrup-spout bottles (Fig. 1C), which consist of a flat, crescent-shaped bottle representing a curved pod with few seeds. The handle of the bottle is a stirrup-like semicircular tube, with a tubular spout at the top. At the base of the spout, there is often a little monkey or other figure. In the Chimu ceramics there are some variants of this type; in some of them the seeds are very bulky, in others, the space they occupy is smooth or scarified, with no traces of seeds.

c) round bottle with stirrup-spout (Fig. 1D). These ceramics consist of a round or almost round bottle with a stirrup-spout handle, as in the preceding type, but with one to three *pacae* pod representations attached to the base of the stirrup. Often, there is a representation of a monkey, complete or only the head, and in some cases, the monkey is replaced by a bird. In a more stylized form, a bottle with two spouts joined by a cylindric handle, the pods are large and flat, with straight grooves representing the depressions between the seeds (Fig. 2A).

d) pacchas. This name is given to the elongated drinking vessels used in ceremonial functions (Carrión, 1955). They were made in the shape of fruits, animals and other objects. In Peruvian pottery, pacchas in the form of *pacae* pods are rather common. In one type, the fruit is slightly curved, with well marked borders, the narrow side smooth; at the basal end, there is a short, wide, cylindrical spout (Fig. 2B). In another type, the pod is almost straight and it has a tubular short spout on top of the upper narrow side, close to the basal end. In Nazca pottery, there is a variant of this type; in place of the spout there is a round vessel, with a band of geometric designs across (Fig. 2C). The Inca pacchas are curved and with a long spout (Fig. 2D).

Historical records

The first written information on *pacae* is a reference of the Spanish historian Pedro Pizarro, in which he states that in 1532 the Inca emperor Atahualpa sent as a gift a basketful of *pacae* to Francisco Pizarro, the conqueror of Peru (Yacovleff & Herrera, 1934). The first description of *pacae* is in the notes of a soldier, Martin Estete written in 1535: "in the plains of Peru...there are fruits of different kinds, such as one called guaba, like cassia fistola but wider...that has white pulp, without seeds, very sweet". Estete uses the term "guaba", derived from guama, a taino word for *Inga* spp. As in the case of maize and other plants, the Spaniards adopted the taino names, spoken only in Hispaniola, spread them all over the continent and took some of them to the Old World. Guaba was later changed to 'caoba', as will be seen in the next paragraph. The term caoba is no longer used for *pacae*; it is widely used in Spanish speaking Latin America for mahogany (*Swietenia* spp.).

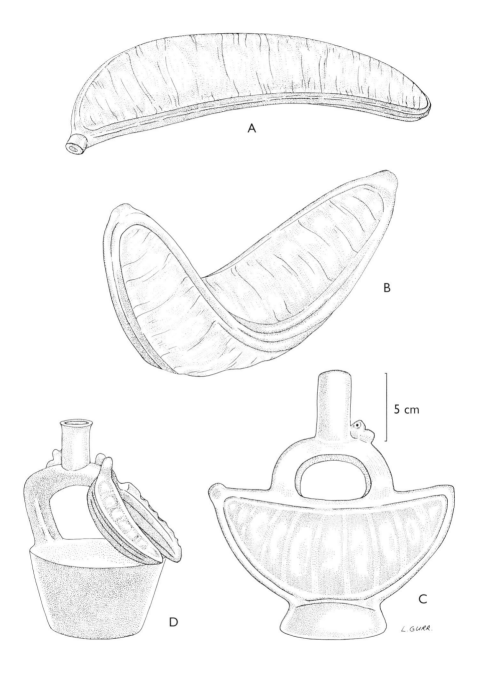

FIG. 1. *Pacae* (*Inga feuillei*) representations in ancient Peruvian pottery. Redrawn from Lavalle (1988) and Martínez (1986) by Linda Gurr.

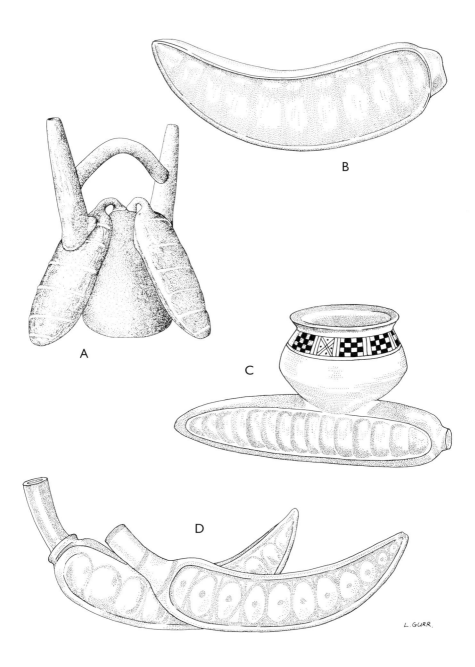

FIG. 2. *Pacae* (*Inga feuillei*) representations in ancient Peruvian pottery. Redrawn from Bird (1962), Carrión (1955), Martínez (1988) and Towle (1961) by the author (A) and Linda Gurr (B–D).

A few years later, another soldier, Pedro Corzo, wrote between 1547–1550, a better description: "There is a tree that produces a fruit two or three palms long, almost as thick as the wrist, that contains inside a part very sweet and juicy; it has a row of separate seeds, like green broad beans, and between them the good tasting substance or fruit. This tree is called caoba, and is large and thick, of hard wood, and the leaves are almost like the service tree".

Oviedo (1959) was not acquainted directly with the *pacae* and his description is based on the information supplied to him by Corzo. He was familiar with the *Inga* species that grow in Hispaniola, and not enthusiastic about their value as fruits: "the Indians used to eat them and the christians too, in case of necessity. I have seen the fruits many times and have tasted them but they seem to me more appropriate for monkeys than for people". The species he knew were *I. vera* and *I. fagifolia* (= *I. laurina*), among the poorest in fruit quality, but he is quite enthusiastic about *pacae.*

Perhaps the best information on *pacae* during the colonial times is in the book by Cobo (1956). "The fruit is a pod, like carob, but varies widely in size according to the place, some no longer than carob, in other, two or three palms long. Between these extremes stand the *pacaes*. The pods of *pacae* are two to three fingers wide and one thick; their rind is stiff and leathery, green outside. In the inside there is a row of seeds as large as broad beans, each one covered by a white substance, spongy and sweet, like cotton coated in sugar. The seeds are dark green, soft and so smooth that at pressing them they slip away". Cobo mentions an interesting use of the pulp, it was dried and eaten as raisins. Also, he mentions the use of the timber in building boats.

The presence of *pacae* in Colombia and Ecuador is listed in the book by Patiño on native fruits of the equatorial region (1963). The *pacae* (*Inga feuillei*) is planted in the coastal valleys of Peru for its fruit, timber, fuelwood and shade, although its use is declining, as noted by Yacovleff & Herrera (1934).

For five centuries, the indigenous fruit trees of South America have been in competition with the Euroasiatic and South East Asian counterparts. Some, like chirimoya and lucuma, have been locally improved in their selection and management. The *pacae* and others have been forgotten, in spite of being very popular, and in the case of *pacae*, possibly the best fruit in the genus.

Since the colonial times, the agriculture of the coastal valleys of Peru has drastically changed several times. The plantation systems for the production of cotton, sugar cane, rice, corn for forage and other crops, have considerably reduced the production of the minor crops. The increased use of land for urbanization and industry has also contributed to that reduction. *Pacae* trees still survive in the coastal region in small numbers around chacras or in the backyards of the suburbs and along irrigation canals, and in the smallholdings in the valleys of the cordilleras. There is an urgent need to collect and evaluate these genetic resources, and to distribute planting material of the superior genotypes. In Peru and other Andean countries, the study of the crops of the high mountains has led in the last three decades to considerable advances in their genetic and agronomic improvement. The results are being reflected in the advancement of the agricultural production of the high Andes. Something similar could be done with the coastal crops; in plants like *pacae*, it may lead to more production of superior types for a market that is always open to the consumption of native crops.

REFERENCES

Bird, J.C. 1962. Art and life in Old Peru: an exhibition. American Museum of Natural History, New York.

Cabrera, L. 1978. Diccionario de aztequismos. D.F., Oasis, Mexico.

Carrión Cachot, R. 1955. El culto al agua en el antiguo Perú. Revista Mus. Nac., Lima 11: 9–100.

Cavalcante, P.B. 1988. Frutas comestiveis de Amazonia. Museo Paraense Emilio Goeldi, Belém, Pará.

Cobo, B. 1956. Obras. Real Academia Espanola, Biblioteca de Autores Españoles, Madrid.

Collazos, Ch. C.; White, P.L.; White, H.S.; Vinas T., E.; Alvistur J., E.; Urquieta A., R.; Vásquez G., J.; Diaz T., C.; Quiros M., A.; Roca N.D., A.; Hegsted, M. & Bradfield, R.B. 1957. La composición de los alimentos peruanos. Ministerio de Salud Pública, Lima.

Dressler, R. L. 1953. The precolumbian cultivated plants of Mexico. Bot. Mus. Leafl. 16: 115–173.

Ducke, A. 1946. Plantas de cultura Precolombina na Amazonia Brasileira. Instituto Agronómico do Norte, Belém, Pará.

Fernandez de Oviedo, G. 1959. Historia general y natural de las Indias. Real Academia Española, Biblioteca de Autores Españoles, Madrid.

Harms, H. 1922. Ubersicht der bisher in altperuanischen grabern gefundenen pflanzenreste. Festschrift Eduard Seler: 157–186. Stuttgart.

Hernandez, F. 1959. Historia natural de la Nueva España. Universidad Nacional de México, Mexico D.F.

Hoehne, F.C. 1948. Frutas indigenas. Instituto de Botanica, Sao Paulo.

Lanning, E.P. 1967. Peru before the Incas. Prentice Hall, Englewood Cliffs, N.J.

Lavalle, J.A. de, ed. 1988. Culturas precolombinas. Chimu. Banco de Crédito de Perú, Lima.

Lumbreras, L. 1974. Los orígenes de la civilización en el Perú. Milla Batres, Lima.

Luna Cavazos, M. 1990. Plantas cultivadas nativas de México. In: T. Rojas (ed.), La agricultura en tierras mexicanas desde sus orígenes hasta nuestros dias. Grijalbo, Mexico D.F.

Martínez, C. 1986. Cerámica prehispanica norperuana. BAR International Series 323(2). Oxford.

National Research Council. 1989. The lost crops of the Incas. National Academy Press, Washington, D.C.

Oviedo, G.F. de 1959, see Fernandez de Oviedo.

Patiño, V.M. 1963. Plantas cultivadas y animales domésticos de América Equinoccial. Imprenta Departamental, Cali, Colombia.

Pearsall, D. 1972. The origins of plant cultivation in South America. In: C.W. Cowan & P.J. Watson, The origins of agriculture. Smithsonian Institution Press, Washington, D.C.

Pennington, T.D. 1997 The genus *Inga* – Botany. Royal Botanic Gardens, Kew.

Rochebrune, M.A.T. 1879. Recherches d'ethnographie botanique sur la flore de sépultures péruviannes d'Ancon. Actes Soc. Linn. Bourdeaux 33(1): 343–358.

Rodriguez Mattos, J. (1978). Frutas indigenas comestiveis do Rio Grande do Sul. IPRNR. Porto Alegre, Brazil.

Safford, W.E. 1917. Food plants and textiles in ancient America. Proceedings of the 19th International Congress of Americanists 12–30. Washington, D.C.

Standley, P.C. & Steyermark, J.A. 1946. Flora of Guatemala, vol. 5. Chicago Natural History Museum, Chicago.

Towle, M.A. 1961. The ethnobotany of pre-colombian Peru. Aldine, Chicago.

Weberbauer, A. 1945. El mundo vegetal de los Andes peruanos. Ministerio de Agricultura, Lima.

Willey, G.R. 1971. An introduction to American archaeology, South America, Prentice Hall, Englewood Cliffs, N.J.

Wittmack, L. 1888. Die nutpflanzen der alten peruanen. Congres International de Americanistes. Berlin 7: 325–349.

Yacovleff, E. & Herrera, F.L. 1934–35. El mundo vegetal de los antiguos peruanos. Revista Mus. Nac., Lima 3: 241–322; 4: 29–102.

CHAPTER 2. GROWTH AND BIOMASS PRODUCTION OF *INGA* SPECIES

T.D. PENNINGTON

INTRODUCTION

Studies on the growth characteristics of *Inga* have until now been largely confined to a single species, *I. edulis* (e.g. Arkoll, 1984), and although it is not stated, the information on this species comes only from its cultivated forms. The origin of the cultivated forms of *I. edulis* is uncertain, though probably Amazonian. It is now widely grown throughout tropical America for its edible fruit, and more recently has attracted attention in agroforestry because of its rapid growth on poor acid soils.

It is now known that there are around 300 species of *Inga* in the Neotropics (Pennington, 1997). *Inga edulis* itself is one of a group of about 10 closely related species in section *Inga*, many of which have similar characteristics of growth and habit. This study is a first attempt to investigate some of the characteristics of the wider genus. Preliminary growth figures for some of these species were reported in Lawrence *et al.* (1994, 1995).

The trial data presented here were collected as part of a wide-ranging survey of the resources of the genus *Inga*, based on a monographic systematic study. The trial work and systematic study took place concurrently from 1991–1996, and the choice of species used in the trials, although not entirely haphazard, was less focused than it would have been if the botanical information now available had been available when designing the trials. In order to set up the trials during a short period in late 1992, it was necessary to use whatever species were in fruit at that time, and which could provide sufficient seed (c. 400 seeds of each species for each trial). Fruit set in *Inga* is not strictly synchronous, the fruit is quickly taken by dispersing animals and predators as soon as it matures, and the seeds have very short viability. In the event seed was generally collected from several individuals of each species.

The objective of the trials was to find *Inga* species with growth characteristics which suggest suitability for forestry or agroforestry use, such as land reclamation, alley-cropping systems and fuelwood. In this context the characteristics of greatest interest are fast growth rates, high production of leafy and woody biomass, ability to establish and flourish on marginal sites and ability to compete well with fast growing secondary vegetation. It must be emphasized that *Inga* species are not true pioneers and because of the nature of their fleshy seeds, are unable to establish on bare earth where exposure to full sun, low humidity and high temperatures will quickly kill them. The most successful species with the above growth characteristics can be broadly described as non-pioneer light-demanding gap species, which are able to exploit the high light conditions of gaps in the forest canopy with their rapid growth, and which eventually become a component of the canopy of the mature forest. *Inga* species of this type tend to be concentrated in two sections of the genus, *Pseudinga* and *Inga* (Pennington, 1997). The least successful species, in terms of the above characteristics are those shade tolerant plants from the forest understorey, which occupy the lower or middle storey of undisturbed forest. These species are slow growing and unable to compete with weedy vegetation.

Trial sites and species

Results are presented here from 3 lowland trial sites in Honduras, Pacific coastal Ecuador, and in Amazonian Ecuador. Full details of their location and the provenance of their component species are to be found in Appendix 1. Wherever possible, local species growing in the vicinity of the trial sites were used, but when locally growing seed was not available, then it was imported from other areas.

Trial design

The trial design was Randomized Complete Blocks arranged to minimize any obvious within-site variation, such as slope. The blocks were replicated 4 times and each contained 4 or 5 species arranged in plots of 48 trees planted at 3 m × 3 m spacing. Full information of the trial design and procedure can be found in Pennington (1994). After the first year's growth each plot was subdivided into a coppice subplot of 24 trees and a shade subplot of 24 trees. The trees in the coppice subplot were pruned at approximately six month intervals at a height of 1.25 m above ground in order to measure branch and leaf biomass production. In order to eliminate edge effects only the central 8 trees of each subplot are used for the analyses. The trees in the shade subplot were left unpruned until they were evaluated for total wood biomass at the end of 3 years.

Each block contained a control plot without *Inga* trees, and the effect of *Inga* shade on weed growth in the shade subplots was measured at approximately 6 month intervals by comparison with the weed growth in the controls. Growth measurements (height, diameters at 30 cm above ground) were taken at approximately 6 month intervals according to the method of Stewart (1990).

RESULTS AND ANALYSES

Growth rates

Heights and diameters measured at 6 month intervals are shown for the Lancetilla, Honduras trial in Figs. 1 and 3, and for the Jatun Sacha, Ecuador trial in Figs. 2 and 4. In the replicated trials *I. edulis* recorded the greatest height (15 m) and diameter (15.5 cm) of all species after 3 years' growth, but only just, as it is closely approached by the related *I. oerstediana* (Lancetilla provenance) with a height of 14 m and diameter of 14.8 cm. In terms of diameter growth most species fall within the narrow range of 13–15.5 cm, with only *I. sapindoides* in the lower range of 8.7–9.5 cm. Different provenances show clear differences when grown on the same site. Two provenances of *I. oerstediana* (Lancetilla and Yojoa) were grown at the Lancetilla trial, where they have almost identical diameters at 3 years but a

FIG. 1. Height growth rates at Lancetilla, Honduras (m).

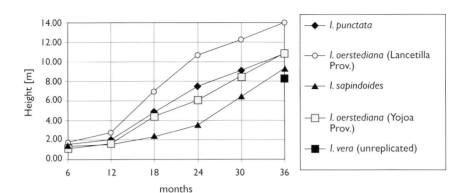

FIG. 2. Height growth rates at Jatun Sacha, Ecuador (m).

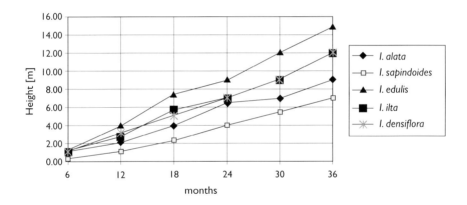

rather large height difference. This contrasts with the situation after 6 months' growth when there was a clear diameter difference. A single unreplicated plot of *I. vera* produced the greatest diameter measurement of all (16.4 cm) at three years. (Fig. 3).

Provenance performance on different sites

A single provenance of *I. edulis* and of *I. alata* were compared at Jatun Sacha in Amazonian Ecuador, and at Rio Pitzará on the Pacific coast of Ecuador. The provenance used originated in Amazonian Ecuador (see Appendix 1). The comparative growth figures are shown in Figs. 5 and 6. In comparison with their performance on the local (Amazonian) site, the

Fɪɢ. 3. Diameter growth rates at Lancetilla, Honduras (cm).

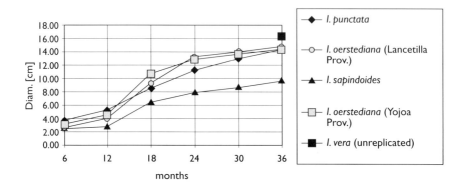

Fɪɢ. 4. Diameter growth rates at Jatun Sacha, Ecuador (cm).

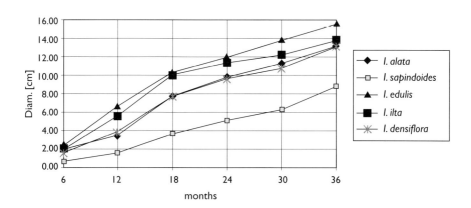

growth of both *I. edulis* and *I. alata* was depressed on the Pacific coastal site. The Pacific site has a slightly lower annual rainfall and a dry season of about 5 months duration, whereas the climate of the Amazonian site is non seasonal. Figs. 5 and 6 also include growth data for two other local *Inga* species (*I. oerstediana* and *I. silanchensis*) which were growing around the Pacific trial site. Both these local species have growth rates which are comparable to or exceed those of the provenances imported from Amazonia. The importance of investigating the utility of local species and provenances, which are well adapted to local climatic, edaphic and other conditions, can not be stressed too highly. Selection of local species of *Inga* is particularly easy, as there are always several co-occurring species to be found at any one

Fig. 5. Height growth rates of *I. edulis* and *I. alata* on 2 sites in Ecuador (m).

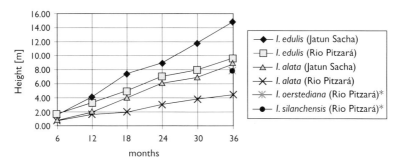

* data from unreplicated plot

Fig. 6. Diameter growth rates of *I. edulis* and *I. alata* on 2 sites in Ecuador (cm), measured at 30 cm above ground level.

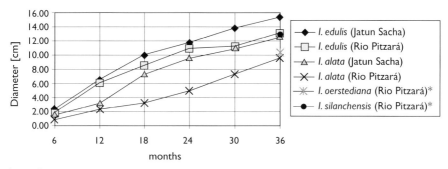

* data from unreplicated plot

place, and many species such as *I. oerstediana* and *I. punctata* are geographically widespread and contain local variants well adapted to widely differing environmental conditions. On the other hand the performance of some other species originating from geographically distant provenances proved in some cases to be very successful. For example, *I. punctata* performed very well at the Lancetilla site, although its seed was derived from locations far removed from the trial site. The trial site at Lancetilla is at sea level on the wet north coast of Honduras, whereas the seed of *I. punctata* originated at an altitude of 1200 m in the much drier south of the country. Similarly *I. vera* performed well at Lancetilla, although it does not appear to be native there.

FIG. 7. Weed control by *Inga* shade (% of unshaded control). (Lancetilla, Honduras).

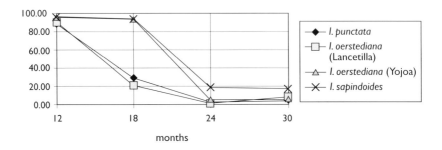

FIG. 8. Weed control by *Inga* shade (% of unshaded control). (Jatun Sacha, Ecuador).

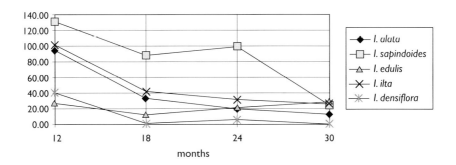

Weed control under Inga shade

Vast areas of lowland tropical America are now covered by abandoned pasture, often a dense mixture of rank introduced grasses, such as *Brachiaria*, intermixed with fast-growing trees and shrubs, such as *Cecropia*, various *Urticaceae, Compositae* and others. The ability of *Inga* species to grow rapidly on marginal lands such as these, combined with its nitrogen fixing properties and production of year-round leaf mulch indicate its potential as a land restorer. In order to quantify one aspect of this potential, the effect of *Inga* shade on the growth of secondary vegetation was measured. Figures 7 and 8 show the effect of *Inga* shade in the uncoppiced subplots on the growth of a mixture of old pasture (*Brachiaria* sp.) and secondary vegetation. Weed growth in the shade subplots is expressed as a percentage of the weed growth in the control plots without *Inga*. Good weed suppression occurs under several species, but there are significant differences between them. Both *I. ilta* and *I. edulis* have a powerful effect within 12 months, and most other species within 18 months. However, after that time species with the greatest vertical growth such as *I. edulis* were less effective than species with less vertical growth and a more profusely branched habit such as *I. ilta* and *I. punctata*. The latter maintained almost total weed control until the end of the trial period, by shading and the production of a thick long-lasting leaf mulch.

Biomass production from coppicing

After the first 12 months' growth, all trees in the coppice subplots were pruned at approximately 1.25 m above ground. All species responded well to coppicing at this height and there was no mortality, as long as a few leafy shoots were left on each plant at the time of pruning. This process was repeated at approximately six month intervals. The productivity from the coppice growth would have been greater if there had been wider spacing between plots and between blocks, as by 24 months there was extensive shading of the coppice subplot by the shade (non-coppiced) subplots, and also by adjoining blocks. For this reason the measurements were abandoned at 30 months. Figs. 9 and 10 show the branch and leaf biomass measured at each coppicing, and Fig. 11 the totals converted to productivity per hectare. Leaf biomass productivity ranges from 4.2 to 7.8 tonnes per hectare per year, and branch biomass from 3 to 10 tonnes per hectare per year. Even at these low density plantings (3 m × 3 m) the productivity is generally very high and compares well with the 17.2 tonnes/hectare/year reported by Hands (1995) for a high density alley cropping experiment. The productivity is much greater than that estimated from the first year's growth of these trials (Lawrence *et al.*, 1994).

Total Wood Biomass

There is much anecdotal evidence that the wood of various *Inga* species is preferred as a source of fuelwood for cooking. The reasons usually cited for this preference are its fast growth, ability to coppice, and the fact that it burns well without much smoke. In order to test the productivity of each species the total wood biomass was measured at the end of the 3 year period. A sample of 3 trees (1 small, 1 medium, 1 large) from each shade subplot was felled and weighed (all wood, including branches down to 2 cm diameter), and the mean biomass per tree and biomass of each species per hectare calculated. The results are expressed in Tables 1 and 2. The most productive species produce between 18.6 and 25 tonnes (dry weight) per hectare per year, at a spacing of 3 m × 3 m. The best species are *I. ilta* (18.6 t), *I. oerstediana* (Lancetilla) (20.4 t), *I. punctata* (21.2 t), *I. vera* (21.9 t), *I. edulis* (24.9 t). These figures are higher than those reported elsewhere (e.g. Arkoll, 1984; Salazar, 1985) though direct comparisons are difficult due to different spacing and management and the different age of the trees. (See also Murphy & Yau, this volume).

Fig. 9. Branch biomass from coppicing.

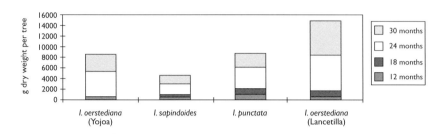

Lancetilla, Honduras

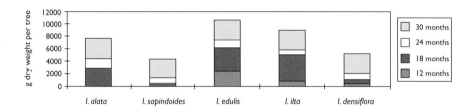

Jatun Sacha, Ecuador

Discussion

Inga is unique among woody legumes in having a wide range of species exhibiting traits useful in one or more areas of agroforestry practice. In the trials described in this chapter, and in Hands chapter, this volume, at least 12 species have been shown to have excellent performance in terms of growth rates, ability to establish and flourish on poor acid soils, competitive ability, production of woody or green biomass, all of which would be of use in some aspects of agroforestry. Where comparisons are available with other (non-*Inga*) species, such as *Erythrina* and *Gliricidia* (see Hands, this volume) *Inga* consistently outperforms them in growth and biomass production and, in addition, has other advantageous characteristics (edible fruit, competitive ability).

Up until now *I. edulis* is the only species which has attracted the interest of agroforesters but although its fast growth and tolerance of poor soils renders it very useful, it does have disadvantages. The cultivated plants of *I. edulis* used for this work are of unknown provenance, probably with a small genetic base (judging from their small morphological variability) and have been protected and spread by man because of their edible fruit and not for any

22

FIG. 10. Leaf biomass from coppice.

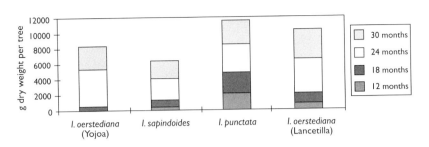

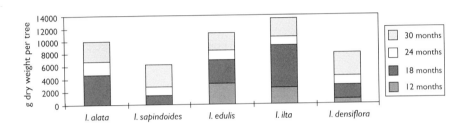

agroforestry merit. Unless it is being grown specifically for this purpose the attractive fruit may be a disadvantage in agroforestry terms (physical damage to crops or trees grown between the *Inga* trees, and *Inga* fruit may be an alternative host for fruit fly (see Ackerman *et al.*, this volume)). *Inga edulis* is also susceptible to attack by mistletoe (*Struthanthus leptostachyus*), which is a strong disincentive to its use in forestry as a nurse crop for timber trees.

The trials described here show that the range of useful traits is not confined to one species nor indeed to just one section of the genus, but extends right across the genus. The sections of the genus which are already know to contain useful species are:

1. *Bourgonia – I. marginata, I. samanensis, I. alata.*

2. *Pseudinga – I. silanchensis, I. densiflora, I. punctata, I. ilta.*

3. *Inga – I. adenophylla, I. edulis, I. oerstediana, I. vera.*

4. *Tetragonae – I. sapindoides.*

FIG. 11. Total biomass from coppice.

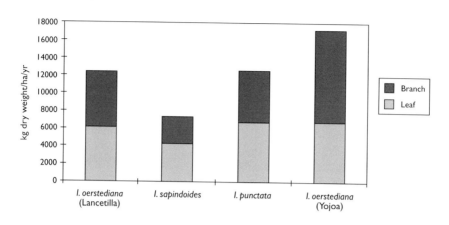

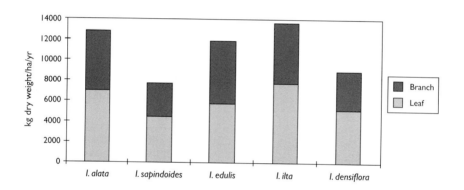

Personal observations by the author suggest that many other species will be added to the list in the coming years.

Although a few 'imported' provenances or species did well on sites where they were not native or naturalized, in general the best results were found where local species were used. For example, local provenances of both *I. marginata* and *I. edulis* performed exceptionally well when planted in extremely poor, highly compacted, acid soil in Amazonian Ecuador. Several species were planted along the oil pipeline road in the Yasuni National Park during 1993–1994. Both species had high survival rates and attained a size of 4 m at two years after planting in soil that consisted of subsoil very low in nutrients and compacted by heavy machinery. Growth rates of non-*Inga* species were much slower and mortality much higher than for *I. edulis* and *I. marginata*

TABLE 1. Total wood biomass (all wood down to 2 cm diameter, kg dry weight), measured after 3 years growth. Lancetilla, Honduras.

	kg/tree	kg/ha/yr
I. punctata	58.5	21235
I. oerstediana (Lancetilla Prov.)	56.2	20400
I. sapindoides	31.9	11580
I. oerstediana (Yojoa Prov.)	44.6	16190
I. vera*	60.6	21998

* unreplicated

TABLE 2. Total wood biomass (all wood down to 2 cm diameter, kg dry weight), measured after 3 years growth. Jatun Sacha, Ecuador.

	kg/tree	kg/ha/yr
I. edulis	68.8	24974
I. alata	37.4	13576
I. ilta	51.3	18622
I. sapindoides	22.4	8131
I. densiflora	27.2	9874

(Neill, pers. comm.). Another local provenance of *I. marginata* gave good results in alley cropping experiments in Costa Rica (Hands, this volume). However, Amazonian provenance *I. marginata* failed in trials on the Pacific coast of Ecuador and *I. marginata* did not establish well in Costa Rica when removed from its native area (Hands, pers. comm). Similarly, Amazonian Ecuador provenance *I. edulis* gave relatively poor results when tested on the Pacific coast of Ecuador. In the latter case, local races of 2 other species (*I. silanchensis* and *I. oerstediana*) performed well. On the other hand some species, such as *I. punctata* and *I. vera* did give good results outside their native area.

The results of the trials described in this volume indicate that agroforesters should be encouraged to investigate the performance of some of the locally growing species as well as using those few species of proven ability.

ACKNOWLEDGEMENTS

This study was financed by the British Government Overseas Development Administration, Projects R.4729 and R.6075 of their Forestry Research Programme. Thanks are due to the following for assistance with setting up and maintaining the trials: A. Lawrence, M. Mijas, F. Montenegro, D. Neill, G. Pilz, N. Revelo, L. Veloz, J. Zuleta, R. Zúniga.

REFERENCES

Arkoll, D.B. 1984. A comparison of some fast growing species suitable for woodlots in the wet tropics. Pesq. Agropecu. Brasil 19: 60–61.

Hands, M.R. 1995. Phosphorus dynamics in slash and burn and alley cropping systems on ultisols in the humid tropics: options for management. In: H. Tiessen (ed.) Phosphorus dynamics in terrestrial and aquatic ecosystems: a global perspective. Proceedings of SCOPE workshop. John Wiley & Sons.

Lawrence, A., Pennington, T.D., Zúniga, R.A., Mijas, M. & Zuleta, J. 1994. Early growth of *Inga* species in Central & South America. Nitrogen Fixing Tree Res. Rep. 12: 74–79.

Lawrence, A., Pennington, T.D., Hands, M.R. & Zúniga, R.A. 1995. *Inga*: high diversity in the Neotropics. In: Nitrogen Fixing Trees for Acid Soils. Proceedings of a workshop held July 3–8, 1994. Turrialba, Costa Rica. Published by the Nitrogen Fixing Tree Association (NFTA), Bangkok, Thailand.

Pennington, T.D. 1994. ODA Research Project R.4729. Study of the genetic resources of the genus *Inga* in Central & South America. Final Report.

Pennington, T.D. 1997. The genus *Inga*: Botany. Royal Botanic Gardens, Kew.

Salazar, R. 1985. Producción de leña y biomasa de *Inga densiflora* Benth. en San Ramon, Costa Rica. Silvoenergia 3. CATIE, Costa Rica.

Stewart, J. 1990. Evaluation Manual, OFI international trial of dry zone hardwood species. ODA, Oxford.

APPENDIX 1

Trial site 1. Honduras, Atlántida, Lancetilla Botanic Garden, Lancetilla, 15°45'N, 87°28'W, altitude 60 m.

Site description: Young secondary regrowth on land formerly covered by mixed broadleaf evergreen tropical forest; level land periodically flooded, sandy loam, low s.o.m., high P, low N, pH 5.2. Rainfall 2950 mm with 2–3 month dry season, mean temperature 25°C.

Species	Seed Provenance			
	Locality	Long./Lat.	Altitude	Habitat type
Inga oerstediana (Lancetilla prov.)	Honduras, Atlántida, Lancetilla	15°45'N, 87°28'W	60 m	secondary forest
I. oerstediana (Yojoa prov.)	Honduras, Comayagua, Lago Yojoa	14°55'N, 87°59'W	750 m	secondary forest
I. punctata	Honduras, Morazan, Monte Carmelo	14°10'N, 87°02'W	1200 m	coffee plantation shade trees
I. sapindoides	Honduras, Atlántida, Lancetilla, Miramar	15°42'N, 87°30'W	230 m	disturbed forest
I. vera	Honduras, Olancho, Rio Patuca, W of Las Plachas	14°35'N, 85°20'W	420 m	riverside

Trial site 2. Ecuador, Napo, Jatun Sacha, 1°03'S, 77°36'W, altitude 400 m.

Site description: Former cacao plantation, originally covered by mixed tropical evergreen forest, on alluvial terrace, level site, non-flooded, clay loam, low-medium s.o.m., very low P, low-medium N, pH 5.1–5.9. Rainfall 4100 mm, uniformly distributed with no dry season, mean temperature 24°C.

Species	Seed Provenance			
	Locality	Long./Lat.	Altitude	Habitat type
Inga alata	Ecuador, Napo, Jatun Sacha	1°03'S, 77°36'W	400 m	old secondary forest
I. densiflora	Ecuador, Napo, Jatun Sacha	1°03'S, 77°36'W	400 m	cultivated trees in gardens
I. edulis	Ecuador, Napo, Tena	1°00'S, 77°49'W	350–400 m	cultivated trees in gardens
I. ilta	Ecuador, road Puerto Napo to Jatun Sacha km 6	1°03'S, 77°50'W	350 m	cultivated trees in gardens
I. sapindoides	Ecuador, Napo, Jatun Sacha	1°03'S, 77°36'W	350–400 m	old secondary forest

Trial site 3. Ecuador, Pichincha, Rio Pitzará, 00°15'N, 79°05'W, altitude 350 m.

Site description: Logged mixed tropical evergreen forest, undulating site, non-flooded, clay loam, medium s.o.m., very low P, low-medium N, pH 4.4–6. Rainfall 3500 mm, seasonal with 5 month dry season, mean temperature 24°C.

Species	Seed Provenance			
	Locality	Long./Lat.	Altitude	Habitat type
Inga alata	Ecuador, Napo, Jatun Sacha	1°03'S, 77°36'W	400 m	old secondary forest
I. edulis	Ecuador, Napo, Tena	1°00'S, 77°49'W	350–400 m	cultivated trees in gardens
I. oerstediana	Ecuador, Pichincha, Rio Pitzará	00°15'N, 79°05'W	350 m	old secondary forest
I. silanchensis	Ecuador, Pichincha, Rio Pitzará	00°15'N, 79°05'W	350 m	disturbed (logged) mixed tropical evergreen forest

CHAPTER 3. CALORIFIC VALUE, BASIC DENSITY AND ASH CONTENT OF *INGA* SPECIES

R.J. MURPHY & P.Y. YAU

INTRODUCTION

The wood biomass produced by the *Inga* species under investigation has the potential to provide a useful source of fuelwood. As part of the evaluation of the properties and productivity of *Inga* species at different sites, samples were evaluated for their gross calorific value (high heating value), basic density and ash content. The gross calorific value of wood is the maximum energy that can be derived from it as a fuel source – it is, in effect, a fundamental property of the material. Often, net calorific values or low heating values are quoted for fuels which represent the amount of energy that can be recovered in a more realistic situation, although values vary depending upon factors such as moisture content and assumed combustion efficiencies.

In the present study, determinations were made of high heating value by bomb calorimetry in order to provide an equivalent basis for comparison of the different *Inga* spp.

MATERIALS AND METHODS

Inga samples

Samples of stemwood of approximate diameter 3–6 cm were received from South America in the air dried condition.

Calorific value

The gross calorific value (GCV) of a known weight of finely ground wood flour samples (0.7–0.8 g per sample) was determined using a Gallenkamp CB-370 Ballistic Bomb Calorimeter. Wood samples for conversion to wood flour were taken mid-way between pith and bark. The majority of determinations were carried out in triplicate, the remainder in duplicate. The unit was calibrated against an analytical grade Benzoic acid standard and appropriate blanks. Between each firing the apparatus was allowed to cool and was then cleaned and dried. Sub-samples of wood flour were taken for separate moisture content determination and all gross calorific values are expressed on an oven dry wood basis.

Basic density

The basic density (oven dry weight/swollen volume) was determined by water displacement using saturated, wedge shaped cross sectional samples (to maintain the proportions of inner and outer wood) weighing approximately 15–25 g. Pith and bark were removed. Samples were saturated by vacuum impregnation with tap water, swollen volume determined by water displacement and then the oven dry weight determined after 24 hours drying at 103°C. Basic density was calculated in Kg/m^3.

Ash content

The content of ash remaining after full combustion (550°C for 3 hours) of 1 g samples of wood flour was determined by weighing.

RESULTS AND DISCUSSION

The data are expressed graphically in Figs. 1–6 and summarized in Table 1. Figs. 1–3 give the data for basic density, GCV and ash content for *Inga* species from Ecuador and Figs. 4–6 for *Inga* species from Honduras.

Fig. 1 illustrates a wide spread in basic density from a low value of 151 Kg/m³ for one sample of *I. oerstediana* to the highest value of 639 Kg/m³ which was for *I. alba*. This is clearly a significant spread of density. However, the majority of values fell in a range of approximately 270–430 Kg/m³ with relatively minor differences being found with location (see *I. edulis* (Jatun Sacha (Amazonian) vs Pitzará (coastal)) and *I. alata* at the same locations).

The gross calorific values show a high degree of consistency between different species and sites. The only exceptions to this were the relatively low values for *I. oerstediana* and *I. carinata* although the former was from an exceptionally low density sample and in both cases the results were derived from single trees making generalisation impossible. The ash content of *Inga* species from the sites varied quite widely from approximately 0.7% to 1.8% by weight.

No consistent relationship was found between GCV, density and ash content. However, the GCVs for *Inga* in these trials are in the expected range for wood (observed spread of values 15 to 25 MJ/Kg (Anderson & Tillman, 1977)) and are in the upper range. The values for *Inga* are similar to those recorded recently for several 'miombo' woodland species of the genera *Brachystegia*, *Terminalia* and *Pterocarpus* in Tanzania (23–25 MJ/Kg) (Keeble, 1997).

A similar picture was found for *Inga* species from Honduras although basic density and GCVs tended to be somewhat higher. Ash content was again highly variable both between and within species.

The data for all species and locations are summarized in Table 1.

When the GCV data are combined with the data for dry wood mass productivity, the potential gross energy production values for a number of the species tested is derived. These values are given in Table 2 and Fig. 7. They provide a useful index for comparison of different species and clearly illustrate the advantages of species such as *I. edulis*, *I. ilta*, *I. punctata*, *I. oerstediana* and *I. vera* for potential energy (fuelwood) production. The values for these *Inga* species are similar to, in some cases higher than, those reported for fast growing *Eucalyptus* species (approx. 440 GJ/ha/yr) and are substantially higher than values for other species such as loblolly pine (168 GJ/ha/yr) or Aspen (63 GJ/ha/yr) (Tillman, 1978).

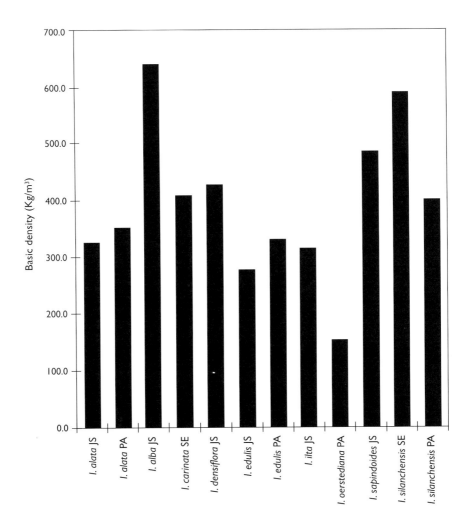

Fig. 1. Basic wood density of Ecuadorian *Inga* species. JS = Jatun Sacha; PA = Pitzará; SE = Silanche.

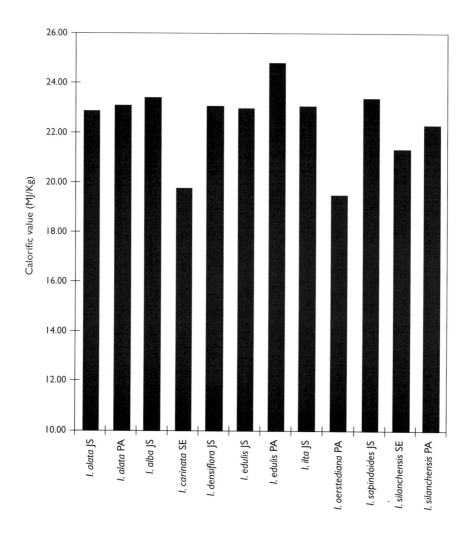

FIG. 2. Gross calorific value of Ecuadorian *Inga* species. JS = Jatun Sacha; PA = Pitzará; SE = Silanche.

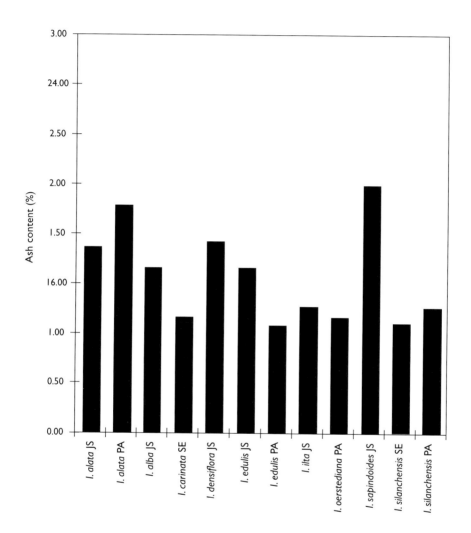

FIG. 3. Ash content of Ecuadorian *Inga* species. JS = Jatun Sacha; PA = Pitzará; SE = Silanche.

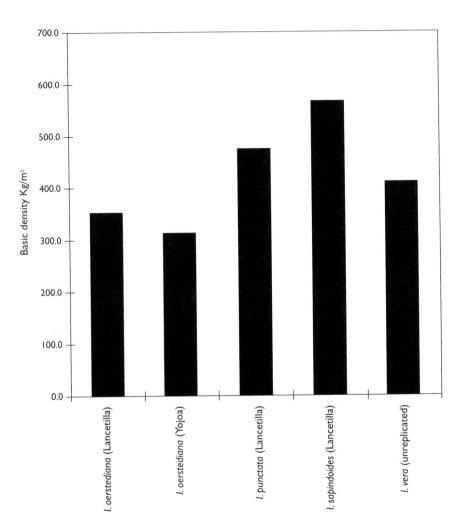

F𝙸𝙶. 4. Basic wood density of Honduran *Inga* species.

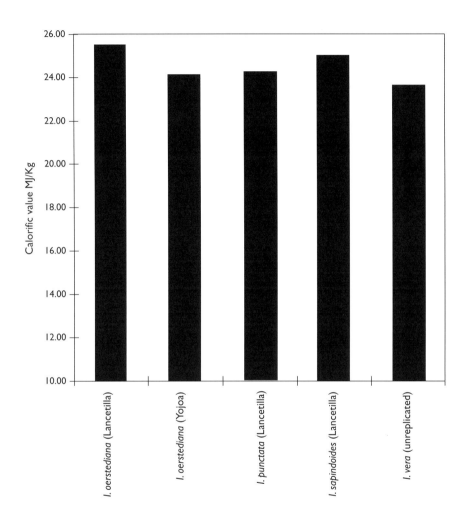

FIG. 5. Gross calorific value of Honduran *Inga* species.

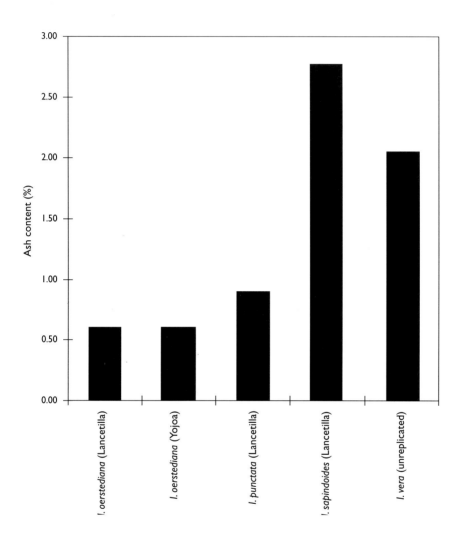

FIG. 6. Ash content of Honduran *Inga* species.

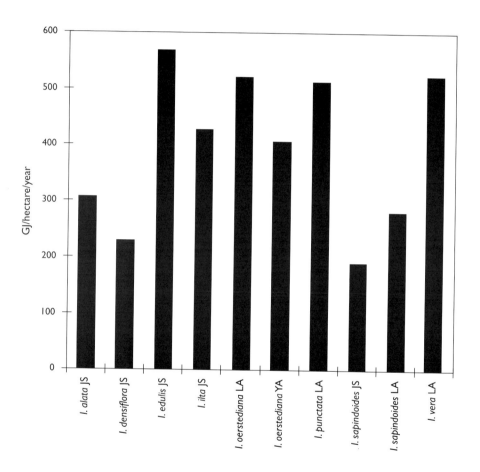

FIG. 7. Potential gross energy value of *Inga* species. JS = Jatun Sacha, Ecuador; LA = Lancetilla, Honduras; YA = Yojoa, Honduras.

TABLE 1. Summary data for *Inga* species.

ECUADOR Species	Region	Basic density (Kg/m³)		Gross calorific value (MJ/Kg)*		Ash content (% dry mass)	
I. alata	Jatun Sacha	**322**	(74)	**22.83**	(2.11)	**1.40**	(0.61)
I. alata	Pitzará	**351**	(66)	**23.07**	(1.65)	**1.70**	(0.48)
I. alba[1,2]	Jatun Sacha	**639**	(n/a)	**23.28**	(n/a)	**1.22**	(n/a)
I. carinata[1]	Silanche	**410**	(n/a)	**19.67**	(n/a)	**0.86**	(n/a)
I. densiflora	Jatun Sacha	**424**	(54)	**22.98**	(2.18)	**1.44**	(0.87)
I. edulis	Jatun Sacha	**271**	(39)	**22.91**	(3.01)	**0.69**	(0.30)
I. edulis	Pitzará	**328**	(102)	**24.76**	(2.29)	**0.81**	(0.35)
I. ilta	Jatun Sacha	**314**	(14)	**22.99**	(2.58)	**0.94**	(0.35)
I. oerstediana[1]	Pitzará	**151**	(n/a)	**19.43**	(n/a)	**0.87**	(n/a)
I. sapindoides	Jatun Sacha	**478**	(13)	**23.36**	(1.59)	**1.86**	(0.17)
I. silanchensis[1]	Pitzará	**399**	(n/a)	**22.24**	(n/a)	**0.93**	(n/a)
I. silanchensis[1]	Silanche	**590**	(n/a)	**21.25**	(n/a)	**0.81**	(n/a)
HONDURAS Species	Region						
I. oerstediana	Lancetilla	**351**	(49)	**25.52**	(1.91)	**0.60**	(0.22)
I. oerstediana	Yojoa	**566**	(67)	**25.00**	(2.09)	**2.77**	(1.41)
I. punctata	Lancetilla	**475**	(110)	**24.20**	(1.75)	**0.91**	(0.41)
I. sapindoides	Lancetilla	**311**	(40)	**24.16**	(1.06)	**0.59**	(0.11)
I. vera[1]	Lancetilla	**409**	(n/a)	**23.71**	(n/a)	**2.04**	(0.67)

*oven dry mass basis; [1]unreplicated; [2]forest tree; n/a not applicable; ( ) = standard deviation

CONCLUSIONS

The values recorded here for *Inga* species are towards the upper range reported for wood (Harker *et al.*, 1982). Whilst some caution is advisable when comparing results from different series of experiments using different equipment these data suggest that the gross calorific values of *Inga* species after relatively short growth periods are equivalent to those for many wood species. This, combined with the high productivity of several species, indicates a high potential value as an energy and fuelwood source. Several additional factors influence the 'quality' of a fuelwood (e.g. combustion rate, presence of calcium oxalate, moisture content, extractive content) (Prior & Cutler, 1992). On the basis of the present data it is recommended that combustion trials are conducted on selected species (*I. edulis* and *I. ilta* in Ecuador and *I. oerstediana*, *I. punctata* and *I. vera* in Honduras) to establish low heating values and practical fuelwood qualities under appropriate conditions.

TABLE 2. Potential gross energy production value of *Inga* species.

Species	Region	Wood biomass (Kg/ha/year)*	Gross calorific value (MJ/Kg)	Gross energy productivity (GJ/ha/year)
I. alata	Jatun Sacha, Ecuador	13576	22.83	310
I. densiflora	Jatun Sacha, Ecuador	9874	22.98	227
I. edulis	Jatun Sacha, Ecuador	24974	22.91	572
I. ilta	Jatun Sacha, Ecuador	18622	22.99	428
I. oerstediana	Lancetilla, Honduras	20400	25.52	521
I. oerstediana	Yojoa, Honduras	16190	25.00	405
I. punctata	Lancetilla, Honduras	21235	24.20	514
I. sapindoides	Jatun Sacha, Ecuador	8131	23.36	190
I. sapindoides	Lancetilla, Honduras	11580	24.16	280
I. vera	Lancetilla, Honduras	21998	23.71	522

* Dry weight basis

REFERENCES

Anderson, L.L. & Tillman, D.A. 1977. Fuels from waste. Academic Press, New York & London.

Harker, A.P., Sandels, A. & Burley, J. 1982. Calorific values for wood and bark and a bibliography for fuelwood. Tropical Products Institute, G162, 1–20.

Keeble, J.J. 1997. Availability and use of fuelwood by the hosts and refugees of the Ulyankulu Refugee Settlement, Tanzania: A study of the environmental impacts of refugees. B301 Dissertation, Department of Geography, University College London.

Prior, J. & Cutler, D. 1992. Trees to fuel Africa's fires. New Sci., 29/8/1992, 35–39.

Tillman, D.A. 1978. Wood as an energy resource. Academic Press, New York & London.

CHAPTER 4. NODULATION AND NITROGEN FIXATION IN THE GENUS *INGA*

E.C.M. FERNANDES

INTRODUCTION

A number of species of *Inga* are used in homegardens, agricultural fields and forest fallows for fruit (Padoch & de Jong, 1991; Lawrence, 1995), as shade trees for perennial crops (Escalante *et al.*, 1987; León, this volume), for fuelwood and for soil conservation and improvement (Staver, 1989). Much of the research on *Inga* has focused on the nutrient cycling potential of *Inga* in managed forest fallows and in agroforestry systems (Palm & Sanchez, 1990; Szott *et al.*, 1991b; Fernandes *et al.*, 1993a; Salazar *et al.*, 1993; Smyth, 1993; Alegre & Rao, 1995; Szott & Palm, 1996). This chapter presents a review of the literature on nodulation and N_2-fixation in *Inga* and the relatively little experimental data on the effects of tree and soil management on nodulation and N_2-fixation. A search of 3.6 million records in the CAB International bibliographic database turned up 260 citations on *Inga*, 11 citations of published studies in which N_2-fixation in *Inga* had been empirically measured and 7 citations about *Inga*'s rhizobial and/or endomycorrhizal symbionts.

NODULATION IN *INGA*

Allen & Allen (1981) cited various authors who reported finding nodules on 13 species of *Inga*. In a more recent survey in the Brazilian Amazon, Moreira *et al.*, (1992) reported nodulation in nine of the twelve species of *Inga* occurring on upland (non-flooded) or flooded areas. The young nodules are usually round, white, and smooth surfaced (Fig. 1). In a study of nodule biomass and acetylene reduction rates in adjacent coffee plantations with shade trees of *I. jinicuil* or *I. vera*, Roskoski (1981) reported that only *I. jinicuil* had nodules and that the coffee production under *I. jinicuil* was 37% higher than under *I. vera*. Mature nodules are brown and rough-surfaced (Allen & Allen, 1936). Seedlings of *Inga edulis* with large numbers of effective nodules (pink to reddish pink interiors) are common in the acid Ultisols in the Peruvian Amazon and the Oxisols in the Brazilian Amazon (Fernandes, pers. observation).

Symbiotic associations among Inga species, rhizobia, and mycorrhizal fungi

Research has shown that effective associations between many leguminous species and mycorrhizal fungi significantly improve growth and nitrogen fixation relative to non-mycorrhizal plants in P-deficient soils (Herrera *et al.*, 1993). Leguminous species have a high P demand for nodulation and nitrogen fixation and VA mycorrhizal fungi enhance P uptake by the host plant (Waidyanatha *et al.*, 1979; Lynd *et al.*, 1985, Fernandes *et al.*, 1991). The increased uptake of P by mycorrhizal roots is due to reduced distances of

41

FIG. 1. Nodules on a three-month-old *Inga edulis* seedling. The inoculum was made up of crushed nodules from mature *Inga* trees, fine roots, and rhizosphere soil mixed together in water. Fresh *Inga* seeds were soaked in the slurry for 24 hours and then planted in freshly tilled soil in the field (Photo: Erick C.M. Fernandes).

diffusion of P from the soil to the VAM fungal hyphae which extend from the root surface (Abbott & Robson, 1982). Other beneficial plant effects attributed to VAM fungi include increased tolerance to drought (Sieverding, 1986) and protection against nematodes (Smith *et al.*, 1986).

Plant species differ in their growth responses to VAM fungal infection depending upon their demand for P and ability to absorb the element from the soil (Hetrick *et al.*, 1988). In general, species having a low root density or few root hairs are very dependent on mycorrhizae (Crush, 1974). Although leguminous species are generally quite dependent on VAM for good growth on P-deficient soils, the dependency varies with plant species, soil available P level, and species of VA mycorrhizal fungi involved (Howeler *et al.*, 1987).

Rhizobia

The only published data found on the rhizobial partners of any species of *Inga* were reports by Allen & Allen (1939) who confirmed that two strains of rhizobia from *Inga laurina* were effective on a number of species that are nodulated by rhizobia in the cowpea miscellany and de Faria (1995) who isolated and tested two efficient rhizobial strains for *Inga marginata*.

Mycorrhizal Fungi

The mycorrhizal fungi that have been found to be associated with *Inga* belong to the group called endomycorrhizae. There is no evidence that *Inga* forms partnerships with the ectomycorrhizae which are important on plantation forestry genera such as *Pinus* and *Eucalyptus*.

Janos (1975) reported that the endomycorrhizal fungi *Acaulospora* sp. and *Sclerocystis dussii* significantly improved height growth, cotyledon retention and the number and length of leaves of *Inga oerstediana* after 24–36 weeks of growth in sterilized alluvial soil in NE Costa Rica. In a greenhouse experiment to study the effects of native mycorrhizal strains and phosphorus addition on the growth of *Inga edulis* seedlings in an acid soil (pH 4.6), Fernandes (1990), found *Acaulospora spinosa* C. Walker & Trappe, *A. foveata* Trappe & Janos, *Scutellospora heterogama* T.H. Nicolson & Gerd., and *Gigaspora decipiens* I.R. Hall & L.K. Abbott from pot cultures inoculated with surface sterilized, fine roots of *Inga edulis*. The addition of rock phosphate to the soil resulted in increased P content and nodulation of *I. edulis* seedlings. Plants inoculated with mycorrhizae and rhizobia had a significantly higher P content and greater number of nodules than plants inoculated with rhizobia alone. The growth effect of inoculation with mycorrhizae was equivalent to the addition of 30 kg P/ha (Fernandes *et al.*, 1991). The data are consistent with the well-known mycorrhizal effect of improving nodulation by improving the P status of the plant (Hayman, 1986).

In a field study in the Peruvian Amazon, Reategui *et al.* (1995) reported mycorrhizal infections in seven species of *Inga*. The species included *I. pilosula*, *I. macrophylla*, *I. edulis*, *I. cinnamomea*, *I. dumosa*, *I. aff. matthewsiana*, and *I. aff. acreana*. Shoot pruning of *I. edulis* reduced the level of mycorrhizal infection for about four months relative to trees that were not pruned. The authors hypothesized that low levels of mycorrhizal inoculum could have resulted in the delayed reinfection of *Inga* roots.

Nitrogen-Fixation in Inga

Although many species of *Inga* have been reported to nodulate, N_2-fixation has been quantified in only four species of *Inga*. Of the eleven citations encountered for measurements of N_2-fixation in *Inga*, eight citations were on *Inga jinicuil* (Roskoski, 1978, 1981, 1982; van Kessel & Roskoski, 1981, 1983; Roskoski *et al.*, 1982; van Kessel *et al.*, 1983; Roskoski & van Kessel, 1985); one on *Inga laurina* (Yatazawa *et al.*, 1983), one on *I. leptolober* (Roskoski, 1978), and one for an unknown *Inga* species (Carpenter, 1992).

An evaluation of the presence and importance of nitrogen fixation in four coffee production systems: coffee only, coffee plus the leguminous shade trees *I. jinicuil* or *I. vera*, and coffee plus banana and orange trees, revealed that total acetylene reduced was highest in the *I. jinicuil* site, and assuming a 3:1 C_2H_2:N_2 ratio, it was estimated that the *Inga* fixed more than 40 kg/ha N/yr (Roskoski, 1982). The activity was primarily associated with *I. jinicuil* nodules. Apparent fixation in the other three sites was less than 1 kg N/ha/yr. Nitrogen fixed in the *I. jinicuil* site was 53% of the average amount of fertilizer nitrogen applied annually. In another study of the nitrogen-fixing capacity of nine legume species, Roskoski *et al.* (1982) estimated that in one coffee plantation, *I. jinicuil* fixed about 35 kg N2/ha per yr. In the same experiment, 20-year-old stands of *Acacia pennatula* and *Gliricidia sepium* fixed 34 and 13 kg N_2 ha^{-1} yr^{-1}, respectively.

Although *Inga* root nodules have been identified as the primary site of significant N_2-fixation (Roskoski, 1982), there is empirical evidence that shows that lenticels on tree bark can also be sites of N_2-fixation. Yatazawa *et al.* (1983) studied nitrogenase development in warty barks of 12 leguminous (including *Inga laurina*) and 9 non-leguminous trees growing in wet tropical and temperate forests. After incubating the bark for a few days with ambient air containing 0.10 atm C_2H_2, they measured acetylene reducing activity (ARA) of 18 nmol C_2H_2 reduced per gram of bark fresh weight per hour. The results suggest that considerable in situ N_2-fixation can occur in the tree bark of *I. laurina* and other forest species.

SOIL AND PLANT MANAGEMENT FACTORS INFLUENCING NODULATION AND N_2-FIXATION IN *INGA*

Inga species are found on a wide range of basic and acidic soil types (Szott *et al.*, 1995; Pennington, 1997). On high base status soils there are likely to be few soil chemical constraints to nodulation. On acidic soils common in the humid tropics, however, research on a range of crop species has shown that both the growth and nodulation of leguminous species may be constrained by high levels of Al saturation, Mn toxicity (Whelan & Alexander, 1986), low P availability and Ca deficiency (Szott *et al.*, 1991a). Van Kessel & Roskoski (1981) studied the effects of soil nutrients on nodulation and N_2-fixation by *I. jinicuil* in a pot experiment. Nodules appeared 2 to 3 weeks earlier and had a higher biomass in high P soil than in low P soil. Although nodule biomass was positively correlated with available soil P, acetylene reducing activity was positively correlated with Mg and K content of the soil. Soil acidity can negatively effect the rhizobia in the soil, inhibit the process of adsorption of

Rhizobium to the root surface, and reduce the root exudates that are responsible for the expression of nodulation genes (Caetano-Anolles *et al.*, 1989; Richardson *et al.*, 1988).

Many species of *Inga* are important components of a range of agroforestry systems because they produce food, fodder and wood and have the potential to enhance the nitrogen status of the soil. In many agroforestry systems, the *Inga* trees are periodically cut and the leaves and branches used for green manure, mulch and fuel. The excessive and frequent removal of shoots can result in severely reduced regrowth and in some cases the plant dies. The effects of management (such as shoot pruning) of leguminous species on the dynamics of root growth, infection by endomycorrhizal fungi, and nodulation are relatively unknown and constitute one of the fundamental research gaps in agroforestry (Fernandes, 1990).

The Effect of Coppicing or Shoot Pruning *Inga* on Nodulation and N_2-fixation

Studies for a range of herbaceous species have shown that shoot removal causes a reduction in root growth. This reduction can be complete, and take place within a few hours following defoliation, eventually leading to increased root mortality (Parker & Sampson, 1930; Hodgkinson & Becking, 1977). In a laboratory experiment with young plants of a pasture species (*Dactylis glomerata* L.), Davidson & Milthorpe (1966) also reported a marked decline in root respiration, and P uptake after severe defoliation. In pastures, grazing of above-ground biomass results in decreased root biomass (Harris, 1978; Christie, 1981).

Research on the shoot pruning of leguminous tree species and the effects on subsequent shoot regrowth indicate that the greater the shoot biomass removed in pruning the slower the subsequent regrowth of the shoot (Das & Dalvi, 1981; Duguma *et al.*, 1988) presumably due to greater reductions in root growth by the heavier pruning. In general, root nodules which are actively fixing atmospheric N_2 are a sink for carbon assimilated via photosynthesis (Gordon *et al.*, 1986; Kouchi *et al.*, 1986). Enhancement of net photosynthate generally increases nodule activity whereas treatments that decrease photosynthesis (defoliation, shading, lodging) reduce nodule activity significantly (Hardy & Havelka, 1976). Estimates of VAM fungal biomass of between 5 to 17% of host root mass have been reported (Kucey & Paul, 1982). The large biomass of fungus plus the concentrations of lipid within the fungal structures suggest that significant quantities of photosynthate are translocated to the fungus (Cooper & Losel, 1978). Root mortality due to shoot pruning or grazing may result in decreased levels of photosynthate for nodules and VA mycorrhizal fungi, reduced root infection by VA mycorrhizae and thereby reduced P uptake, nodule activity, and regrowth of the entire leguminous host species.

In a study to measure the influence of various intensities (25, 50 and 75 percent) of shoot pruning of *I. edulis* on shoot and root dynamics, root infection by VAM fungi, and nodulation, and N and P accumulation in shoot regrowth, Fernandes (1990), reported that shoot pruning of 3-month-old *Inga edulis* seedlings significantly reduced the biomass of live fine and coarse roots within four weeks following shoot pruning. The greater the intensity of shoot pruning the greater the measured decline in root biomass, VA mycorrhizal infection, and nodule activity. The total root length infected with VAM fungi at 64 days after shoot pruning (DASP) was significantly lower than for controls at both 50% and 75% shoot pruning intensities (Fig. 2). At shoot pruning intensities of between 25% and 50%, *Inga* plants formed new and active nodules between 32 and 64 DASP. A significant decline in the number of active nodules was detected at 8 days after 75% of shoot biomass was pruned (Fig. 3) and the number of active nodules remained significantly lower in the 75% shoot prune treatment relative to the 25% and 50% shoot prune treatments 64 days after the shoots were pruned.

CONCLUSIONS AND RECOMMENDATIONS

There are several management options for improving the growth, N_2-fixation, and productivity of *Inga* species on acid, infertile soils (Fernandes *et al.*, 1993b). Four key interventions will significantly improve the growth and productivity of Inga species in agroforestry systems:

1. Supply modest amounts (20–50 kg ha[-1]) of important nutrients like P, Ca and Mg that are deficient in acid soils. Nutrients exported in crop or biomass harvests need to be replaced to sustain the productivity of the systems and *Inga* species. Fernandes *et al.* (1993a), reported significantly improved biomass production and nutrient content of fertilized relative to non-fertilized *I. edulis* in an alley cropping experiment on an Ultisol in the Peruvian Amazon. Hutton (1981) showed that even for *Leucaena leucocephala*, which grows poorly on acid soils, the source of acid-soil tolerance was based on Ca uptake rather than Al tolerance. Liming to reduce soil acidity, however, is not recommended (Kass, 1995).

2. Promote the tripartite symbiosis among *Inga*, rhizobia and mycorrhizal fungi. Several researchers have reported the enhancement of growth, nodulation and N_2 fixation in leguminous species associated with VA mycorrhizal fungi, especially in P-deficient soils (Herrera *et al.*, 1993; Saif, 1987). Ensuring that seedlings of *Inga* are exposed to adequate levels of inoculum of rhizobia and VAM fungi at an early stage of their development is the first step to improving their growth and productivity in agroforestry systems. An efficient way to achieve effective nodulation and mycorrhizal infection of *Inga edulis* in the field was reported by Fernandes *et al.* (1993a). The inoculum was produced by collecting surface soil, nodules and fine roots from mature, nodulated trees. Freshly collected *Inga* seeds were left to soak for 12 hours in the slurry made from the soil, fine roots and crushed nodules. The seeds were then removed from the slurry and planted into a freshly tilled patch of soil and the slurry was poured evenly over the seed bed. Two months later, around 95 percent of the seedlings were profusely nodulated (Fig. 1) and 85 percent of randomly selected and sectioned nodules had a light reddish pink colour. Cleared and stained fine roots of these nodulated seedlings showed that they were well colonized by VA mycorrhizal fungi. The inoculum of fine

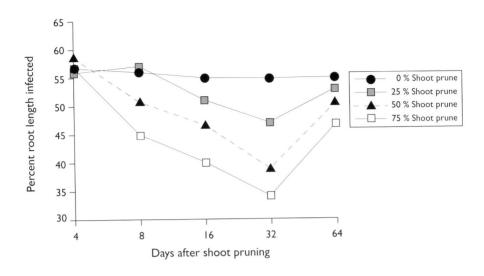

FIG. 2. Percent root length of *Inga edulis* infected with VAM fungi as a function of shoot pruning intensity and time after shoot pruning (Fernandes, 1990).

roots, nodules and rhizosphere soil was effective in providing both the rhizobia and mycorrhizal fungi and the method is suitable for use by farmers under field conditions.

3. Select and use suitable provenances (seed sources) of *Inga* species for each site and for the intended use or function of *Inga* (e.g. mulch, green manure, fuelwood). Szott (1995) reported significant differences in growth rates, biomass productivity and tolerance to pruning among nine provenances of *I. edulis* in the Peruvian Amazon. Similar significant differences in above- and below-ground productivity among provenances were reported for *Gliricidia sepium*, another nitrogen-fixing species (Fernandes, 1990). It is important to mention that Fernandes (1990) also found differences in levels of infection by native VAM fungi among the provenances of *G. sepium*.

4. Select and use suitable rhizobial and mycorrhizal fungi for different species and provenances of *Inga*. In a study of the occurrence of nodulation in 594 leguminous tree species in Brazil, de Faria (1995) reported the isolation of efficient *Rhizobium* strains for a number of nitrogen-fixing legumes. The rhizobial strains isolated from caesalpinoid tree and shrub species appeared to be the most adapted to acid soils. No published reports were found of such selections for endomycorrhizal species for *Inga*. Howeler *et al.* (1987) and Saif (1987) evaluated a range of native VAM fungal strains and reported enhanced growth responses of cassava and a variety of pasture legumes to inoculation with the VAM fungi *Glomus manihotis* and *Entrophospora colombiana* in acid soils.

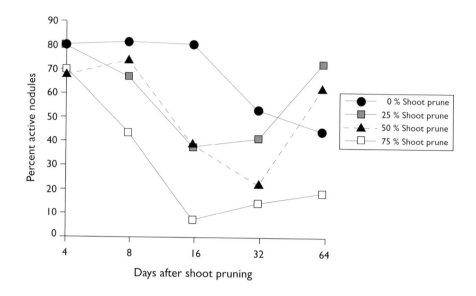

FIG. 3. Percent active nodules in *Inga edulis* as a function of shoot pruning intensity and time after shoot pruning (Fernandes, 1990).

REFERENCES

Abbott, L.K. & Robson, A.D. 1982. The role of vesicular-arbuscular fungi in agriculture and the selection of fungi for inoculation. Austral J. Agric. Res. 33: 389–397.

Alegre, J.C. & Rao, M.R. 1995. Soil and water conservation by contour hedging in the humid tropics of Peru. Agric. Eco-Syst. Environm. 57(1): 17–25.

Allen, O.N. & Allen, E.K. 1936. Root nodule bacteria of some tropical leguminous plants: I. Cross inoculation studies with *Vigna sinesis* L. Soil Sci. 42 (1): 61–77.

Allen, O.N. & Allen, E.K. 1939. Root nodule bacteria of some tropical leguminous plants: II. Cross inoculation tests within the cowpea group. Soil Sci. 47 (1): 63–76.

Allen, O.N. & Allen, E.K. 1981. The *Leguminosae*, a source book of characteristics, uses, and nodulation. Univ. Wisconsin Press. Madison, Wisconsin, USA.

Caetano-Annolles, G., Lagares, A. & Favelukes, G. 1989. Adsorption of *Rhizobium meliloti* to alfalfa roots. Dependence on divalent cations and pH. Pl. & Soil, 117: 67–74.

Carpenter, E.J. 1992. Nitrogen fixation in the epiphyllae and root nodules of trees in the lowland tropical rainforest of Costa Rica. Acta Oecol. 13 (2). 153–160.

Christie, E.K. 1981. Ecosystems processes in semi-arid grasslands. I. Primary production and water use of two communities possessing different photosynthetic pathways. Austral J. Agric. Res. 29: 773–787.

Cooper, K.M. & Losel, D.M. 1978. Lipid physiology of vesicular-arbuscular mycorrhizae. I. Composition of lipids in roots of onion, clover and ryegrass infected with *Glomus mosseae.* New Phytol. 80: 143–151.

Crush, J.R. 1974. Plant growth responses to vesicular-arbuscular mycorrhiza. VII. Growth and nodulation of some herbage legumes. New Phytol. 73: 743–749.

Das, R.B. & Dalvi, G.S. 1981. Effect of interval and intensity of cutting of *Leucaena leucocephala.* Leucaena Res. Rep. 2: 21–22.

Davidson, J.L. & Milthorpe, F.L. 1966. The effect of defoliation on the carbon balance in *Dactylis glomerata.* Ann. Bot. (London) 30: 184–198.

de Faria, S.M. 1995. Occurrence and rhizobial selection for legume trees adapted to acid soils. Pp. 295–300. In: D.O. Evans & L.T. Szott, (eds.), Nitrogen-fixing trees for acid soils. Proceedings of a workshop held July 3–8, 1994. Turrialba, Costa Rica. Published by the Nitrogen Fixing Tree Association (NFTA), Bangkok, Thailand.

Duguma, B., Kang, B.T. & Okali, D.U.U. 1988. Effect of pruning intensities of three woody leguminous species grown in alley cropping with maize and cowpea on an Alfisol. Agroforest. Systems 6: 19–35.

Escalante, F.E.E, Aguilar, R.A., & Lugo, P.R. 1987. Identificación, evaluación y distribuición espacial de especies utilizadas como sombra en sistemas tradicionales de café. Venezuela Forestal. 3(11): 50–62.

Fernandes, E.C.M. 1990. Alley cropping on acid soils. Ph.D. Dissertation. North Carolina State University, Raleigh, NC.

Fernandes, E.C.M., Davey, C.B., & Meléndez, G. 1991. Vesicular arbuscular mycorrhizae and the growth of tropical leguminous tree and pasture legumes in acid soils. Part I: Rock phosphate effects. Pp. 228–230. In: N. Caudle, (ed.), TropSoils Technical Report 1988–89. North Carolina State University, Raleigh, NC, USA.

Fernandes, E.C.M, Davey, C.B. & Nelson, L.A. 1993a. Alley cropping on an Ultisol in the Peruvian Amazon: Mulch, fertilizer and tree root pruning effects. Pp. 77–96. In: J. Ragland & R. Lal, (eds.), Technologies for Sustainable Agriculture in the Tropics. Amer. Soc. Agron. Special Publ. 56. ASA, Madison, WI.

Fernandes, E.C.M., Garrity, D.P., Szott, L.A. & Palm, C.A. 1993b. Use and potential of domesticated trees for soil improvement. In: R.R.B. Leakey & A.C. Newton, (eds.), Tropical Trees: The Potential for Domestication. Proc. IUFRO Centennial Year (1892–1992) Conf. Edinburgh, pp. 218–230. HMSO, London.

Gordon, A.J., Mitchell, D.F., Ryle, G.J.A., & Powell, C.E. 1986. Diurnal production and utilization of photosynthate in nodulated white clover. J. Exp. Bot. 38: 84–98.

Hardy, R.W.F. & Havelka, U.D. 1976. Photosynthate as a major factor limiting nitrogen fixation by field-grown legumes with emphasis on soybeans. Pp. 421–439. In: P.S. Nutman, (ed.), Symbiotic nitrogen fixation in plants. International Biology Program Series. Vol. 7. Cambridge University Press, London.

Harris, W. 1978. Defoliation as a determinant of the growth, persistence and composition of pasture. Pp. 67–83. In: J.R. Wilson, (ed.), Plant relations in pastures. Commonwealth Scientific and Industrial Research Organization, Canberra, Australia.

Hayman, D.S. 1986. Mycorrhizae of nitrogen-fixing legumes. Microbial Resources Centers Journal. 2: 121–145.

Herrera, M.A., Salamanca, J. & Barea J.M. 1993. Mycorrhizal associations and their functions in nodulating nitrogen fixing trees. Pp. 141–158. In: N.S. Subba Rao & C. Rodriguez-Barruecco, (eds.), Symbiosis in nitrogen fixing trees. Oxford & IBH Publishing Co. New Delhi, India.

Hetrick, B.A.D., Kitt, D.G. & Wilson, G.T. 1988. Mycorrhizal dependence and growth habit of warm season and cool season tallgrass prairie plants. Canad. J. Bot. 66: 1376–1380.

Hodgkinson, K.C. & Becking, H.G.B. 1977. Effect of defoliation on root growth of some arid zone perennial plants. Austral. J. Agric. Res. 29: 31–42.

Howeler, R.H., Sieverding, E. & Saif, S. 1987. Practical aspects of mycorrhizal technology in some tropical crops and pastures. Pl. & Soil 100: 249–283.

Hutton, E.M. 1981. Natural crossing and acid-tolerance in some *Leucaena* species. Leucaena Res. Rep. 2: 2–4.

Janos, D. P. 1975. Effects of vesicular-arbuscular mycorrhizae on lowland tropical rainforest trees. Pp. 437–446. In: F.E. Sanders, B. Mosse & P.B. Tinker, (eds.), Endomycorrhizas. Proceedings of a Symposium held at the University of Leeds, 22–25 July 1974. Academic Press. London.

Kass, D.L. 1995. Are nitrogen fixing trees a solution for acid soils? Pp. 19–31. In: D.O. Evans & L.T. Szott, (eds.), Nitrogen-fixing trees for acid soils. Proceedings of a workshop held July 3–8, 1994. Turrialba, Costa Rica. Published by the Nitrogen Fixing Tree Association (NFTA), Bangkok, Thailand.

Kouchi, H., Akao, S. & Yoneyama, T. 1986. Respiratory utilization of ^{13}C-labelled photosynthate in nodulated root systems of soybean plants. J. Exp. Bot. 37: 985–993.

Kucey, R.M.N. & Paul, E.A. 1982. Carbon flow, photosynthesis, and N_2 fixation in nodulated faba beans (*Vivia faba* L.). Soil Biol. Biochem. 14: 407–412.

Lawrence, A. 1995. Farmer knowledge and use of *Inga* species. Pp. 142–151. In: D.O. Evans & L.T. Szott, (eds.), Nitrogen-fixing trees for acid soils. Proceedings of a workshop held July 3–8, 1994. Turrialba, Costa Rica. Published by the Nitrogen Fixing Tree Association (NFTA), Bangkok, Thailand.

Lynd, J.Q., Tyrl, R.J. & Purcino, A.A.C. 1985. Mycorrhiza-soil fertility effects on regrowth, nodulation and nitrogenase activity of siratro (*Macroptilium atropurpureum* (DC) Urb.). J. Pl. Nutr. 8: 1047–1059.

Moreira, F.M.M., Silva, M.F. & Faria, S.M. 1992. Occurrence of nodulation in legume species in the Amazon region of Brazil. New Phytol. 121: 563–570.

Paddoch, C. & de Jong, W. 1991. The house gardens of Santa Rosa: diversity and variability in an Amazonian agricultural system. Econ. Bot. 45(2): 166–175.

Palm, C.A. & Sanchez, P.A. 1990. Decomposition and nutrient release patterns of the leaves of three tropical legumes. Biotropica. 22(4): 330–338.

Parker, K.W. & Sampson, A.W. 1930. Influence of leafage removal on anatomical structure of roots of *Stipa pulchra* and *Bromus hordeus*. Pl. Physiol. (Lancaster) 5: 543–553.

Pennington, T.D. 1997. The genus *Inga*: Botany. The Royal Botanic Gardens, Kew.

Reategui, A.U., Szott, L.T., & Ricse, A. 1995. Mycorrhizal infection in *Inga* species from the Peruvian Amazon. Pp. 301–312. In: D.O. Evans & L.T. Szott, (eds.), Nitrogen-fixing trees for acid soils. Proceedings of a workshop held July 3–8, 1994. Turrialba, Costa Rica. Published by the Nitrogen Fixing Tree Association (NFTA), Bangkok, Thailand.

Richardson, A.E., Djordjevic, M.A., Rolf, B.G. & Simpson, R.J. 1988. Effects of pH, Ca, and Al on the exudation from clover seedlings of compounds that induce the expression of nodulation genes in *Rhizobium trifolii*. Pl. & Soil 109: 37–47.

Roskoski, J.P. 1978. The importance of nitrogen fixation from non-crop legumes in a tropical agro-ecosystem. Bull. Ecol. Soc. Amer. 59 (2): 52.

Roskoski, J.P. 1981. Nodulation and N_2-fixation by *Inga jinicuil*, a woody legume in coffee plantations. I. Measurements of nodule biomass and field C_2H_2 reduction rates. Pl. & Soil 59 (2): 201–206.

Roskoski, J.P. 1982. Nitrogen fixation in a Mexican coffee plantation. Pl. & Soil 67 (1): 283–291.

Roskoski, J.P., Montano, J., van Kessel, C. & Castilleja, G. 1982. Nitrogen fixation by tropical woody legumes: potential source of soil enrichment. In: Biological nitrogen fixation technology for tropical agriculture. 447–454. CIAT, Colombia.

Roskoski, J.P. & van Kessel, C. 1985. Annual, seasonal and diel variation in nitrogen fixing activity by *Inga jinicuil*, a tropical leguminous tree. Oikos. 44 (2): 306–312.

Saif, S.R. 1987. Growth responses of tropical forage plant species to vesicular-arbuscular mycorrhizae. I. Growth, mineral uptake and mycorrhizal dependency. Pl. & Soil 97: 25–35.

Salazar, A., Szott, L.T. & Palm, C.A. 1993. Crop-tree interactions in alley cropping systems on alluvial soils of the upper Amazon basin. Agroforest. Systems 22(1): 67–82.

Sieverding, E. 1986. Influence of soil water regimes on vesicular-arbuscular mycorrhiza: IV. Effect on root growth and water relations of *Sorghum bicolor*. J. Agron. Crop. Sci. 157: 36–42.

Smith, G.S., Hussey, R.S. & Roncadori, R.W. 1986. Penetration and post-infection development of *Meloidogyne incognita* on cotton as affected by *Glomus intraradices* and phosphorus. J. Nematol. 18: 429–435.

Smyth, S. 1993. The role of trees in tropical agroforestry. Ph.D. thesis. Cambridge University, Dept. of Plant Sciences, Cambridge, UK.

Staver, C. 1989. Shortened bush fallow rotations with relay-cropped *Inga edulis* and *Desmodium ovalifolium* in wet central Amazonian Peru. Agroforest. Systems 8: 173–196.

Szott, L.T. 1995. Growth and biomass production of nitrogen fixing trees on acid soils. Pp. 61–76. In: D.O. Evans & L.T. Szott, (eds.), Nitrogen-fixing trees for acid soils. Proceedings of a workshop held July 3–8, 1994. Turrialba, Costa Rica. Published by the Nitrogen Fixing Tree Association (NFTA), Bangkok, Thailand.

Szott, L., Fernandes, E.C.M. & Sanchez, P.A. 1991a. Soil-Plant Interactions in Agroforestry. Agric. Eco-Syst. Environm. 45: 127–152.

Szott, L., Palm, C.A. & Sanchez, P.A. 1991b Agroforestry in acid soils of the humid tropics. Advances Agron. 45: 275–301.

Szott, L.T., Ricse, A. & Alegre, J. 1995. Growth and biomass production by introductions of *Inga edulis* in the Peruvian Amazon. Pp. 237–249. In: D.O. Evans, & L.T. Szott (eds.) Nitrogen-fixing trees for acid soils. Proceedings of a workshop held July 3–8, 1994. Turrialba, Costa Rica. Published by the Nitrogen Fixing Tree Association (NFTA), Bangkok, Thailand.

Szott, L.A. & Palm, C.A. 1996. Nutrient stocks in managed and natural humid tropical fallows. Pl. & Soil 186 (2): 293–309.

Van Kessel, C. & Roskoski, J.P. 1981. Nodulation and N_2 fixation by *Inga jinicuil*, a woody legume in coffee plantations. II. Effect of soil nutrients on nodulation and N_2 fixation. Pl. & Soil 59 (2): 207–215.

Van Kessel, C. & Roskoski, J.P. 1983. Nodulation and N_2 fixation in *Inga jinicuil*, a woody legume in coffee plantations. III. Effect of fertilizers and soil shading on nodulation and nitrogen fixation (acetylene reduction) of *I. jinicuil* seedlings. Pl. & Soil 72 (1): 95–105.

Van Kessel, C., Roskoski, J.P., Wood, T. & Montano, J. 1983. N_2 fixation and H_2 evolution by six species of tropical leguminous trees. Pl. Physiol. (Lancaster) 72 (3): 909–910.

Waidyanatha, U.P. De S., Yogaratnan, N. & Aritaratne, W.A. 1979. Mycorrhizal infection on growth and nitrogen fixation of *Pueraria* and *Stylosanthes* and the uptake of phosphorus from two rock phosphates. New Phytol. 82: 147–152.

Whelan, A.M. & Alexander, M. 1986. Effects of low pH and high Al, Mn, and Fe levels on the survival of *Rhizobium trifolii* and the nodulation of subterranean clover. Pl. & Soil 92: 363–371.

Yatazawa, A.U., Hambali G.G., & Uchino, F. 1983. Nitrogen fixing activity in warty lenticellate tree barks (*Inga laurina*). Soil Sci. Pl. Nutr. 29 (3): 285–294.

CHAPTER 5. THE USES OF *INGA* IN THE ACID SOILS OF THE RAINFOREST ZONE: ALLEY-CROPPING SUSTAINABILITY AND SOIL-REGENERATION

M.R. HANDS

INTRODUCTION

In this chapter, the potential role and uses of *Inga* are set against the extensive and environmentally-damaging failure of traditional shifting cultivation systems to sustain intensification under increased population pressure (Brookfield, 1988). The case of Central America and more recently, Amazonia, is a sufficiently poignant example, but there are many more such examples in all of the world's three great humid tropical regions. In the humid neotropics, this vast forest zone, sparsely populated in historical times at least, has, in recent decades, been opened up to logging; this is often followed by the invasion of slash-and-burn subsistence farmers displaced from elsewhere; and, in South and Central America, at least, by the final conversion of exhausted swidden land to extensive cattle ranching (Schmink & Wood, 1984; Hecht, 1985; Denslow & Padoch, 1988).

Over-intensive swidden agriculture is an essentially consumptive use of temporarily available forest soil fertility. As rural populations grow on marginal lands and as plantation agriculture expands over more fertile river basin soils, so access to fresh forest fallows diminishes. The system is now failing widely in tropical America.

The emphasis here will be placed upon experimental evidence and upon the potential role of *Inga* as a component of low-input, sustainable, subsistence systems of agriculture; and, increasingly, in measures designed to regenerate the fertility of land which has already been degraded by over-use. Low (or zero)-input strategies need to be available to farmers as an alternative to failing swidden systems, and as a means of regenerating those landscapes in which shifting agriculture has already failed. Such strategies must lie within the control of the farmers themselves and must not expose them to debt or dependency upon externalities. If it were the case that they had additional resources, if they could afford to purchase and transport those soil supplements that help to maintain fertility, then so much the better.

In this chapter, the possible role of nitrogen-fixing trees (NFTs) in sustainable cropping systems is outlined in general terms; this is followed by a more detailed treatment as to why the outstanding productivity and acid soil tolerance of *Inga* are potentially so important in this context. A number of the most salient qualities claimed for *Inga* as a coppicing legume in alley-cropping, and as a pioneer regenerator of fertility in degraded soil, are covered in more detail below; they are summarised in Table 1; but much more needs to be known.

TABLE 1. A summary of characteristics of *Inga*; claims and comments.

CLAIMED FOR *INGA*	EVIDENCE OR COMMENT
1. High productivity on, and tolerance of, acid tropical latosols	Commonly observed; also see data below
2. Tolerance of coppicing or pruning	Commonly observed in *Inga* shade over coffee; but manner of pruning can be important
3. Nitrogen fixing ability	Difficult to quantify, but consistent with observed high production of foliage
4. Recalcitrant mulch; physical protection of the uppermost and most important layers of soil from heat and desiccation by the sun and from structural damage, erosion by rain	Observed in a number of studies. Emerging from these studies (see below) as one of the most important characteristics of alley-cropping with *Inga*
5. Apparent resistance to important root pathogens; eg. Root-knot nematodes	See observations from San Juan site, below, re. *Meloidygone spp.*
6. The possible biological protection of intercropped cultivars due to the presence of aggressive ants and other insect predators associated with extra-floral nectaries	Some supporting data and informal observations, but not fully explored. In addition to many ant species, predatory and parasitoid wasps, widely observed on *Inga* at the two sites reported below
7. Weed suppression beneath mulch in alley-cropping or beneath the canopy in free-growing situations	Observed in a number of studies. Strong weed suppression observed at San Juan site (see below)
8. Flexibility: Sufficient diversity between species (and species groups) to enable differing growth and foliage characteristics to be chosen for a specific role	Some broad characteristics observed, but a very small proportion of the total 300 species have been placed under trial
9. Multiple uses; regeneration of degraded soil, formerly under forest and now degraded by exposure, cattle range, etc. *Inga* is also a preferred firewood	Growth form and characteristics of many *Inga* species should lend them to these uses. Early results of trials support this contention

Some of the characteristics listed here are beginning to emerge as being of paramount importance in alley-cropping; they are given in more detail below. Before such detailed treatment, however, it is important to set the context in which the properties of *Inga*, particularly in alley-cropping, can be expected to function. The critical context is taken to be that of failed shifting cultivation systems on the acid latosols of the humid tropics. In other tropical contexts, indigenous agricultures have provided the basis for an intensification capable of feeding a vastly increased human population (e.g. padi rice). On acid soils there are no such examples to provide a starting point; and, today, it is clear that shifting, fallow systems cannot withstand intensification. The world's humid tropics have never before seen the levels of subsistence and economic pressure which bear upon them today and which result in a consumptive and accelerating destruction of these regions' remnant primary and secondary forest cover.

It will be argued below that the only model available to us, of sustainability on zero inputs, is that of the tropical rainforest itself. The special characteristics of "oligotrophic" rainforest ecosystems on acid humid tropical latosols (Jordan, 1985; 1989) are taken here as the unavoidable minimal conditions that will have to be incorporated into any low-input, sustainable alternative to shifting cultivation. Shallow rooting and high dependence on mycotrophism are just two such characteristics.

The soil types in which sustainable alternatives to shifting agriculture will be expected to function were themselves formed under the biological, chemical and climatic conditions of rain-forest. A case is made here that, in simulation of the soil-litter conditions of the forest itself, any alternative agricultural system would have to generate, *in situ*, input levels of organic matter (OM) which are comparable to those of the forest.

Any such system would differ from that of the forest in that (it is assumed) higher levels of biologically fixed nitrogen would be in flux; and that the litter component, being green leguminous mulch, would contain higher quantities of N and P than dead litter. In these respects, and because maintenance levels of soil supplements will almost certainly have to be added from time-to-time, the system is being "driven" in comparison to the forest ecosystem. Finally, the case is made for *Inga* as outstanding acid-soil-tolerant leguminous trees, to be grown in alley configuration.

The tough mulch of *Inga* is seen here as particularly important in, firstly, encouraging shallow-rooting, beneath its protection, and away from the antagonistic, acid mineral soil; and, secondly as a major suppressor of weed growth.

Acronyms used are as follows:

A-C	Alley-cropping
a-g	Above-ground (e.g. above-ground biomass).
b-g	Below-ground.
CEC	Cation Exchange Capacity
CR	Costa Rica
CSJ	Co-operativa San Juan (one of the a-c trial sites in Costa Rica).
DM	Dry-matter (all data refer to oven-dry weights).

E/G   *Erythrina - Gliricidia* alleys.
La C   La Conquista, Sarapiqui, Costa Rica (one of the a-c trial sites).
NFT   Nitrogen-fixing tree.
NPP   Net Primary Production (of an ecosystem or agro-ecosystem).
OM   Organic Matter.
SOM   Soil Organic Matter.
S-C   Shifting cultivation.
VAM   Vesicular-arbuscular Mycorrhiza.

Viable alternatives to traditional shifting cultivation systems are urgently required; not only, as sustainable low-input alternative systems for staple food and cash-crop production; but also, as low-cost strategies for regenerating the fertility and condition of formerly forested soils which are now degraded by years of exposure under swidden and cattle range. The need is for systems to function viably, within the control of the farmer, at least during the transition from shifting to settled forms of subsistence (assuming such forms are possible on the more acid, leached tropical soils) and, if possible, to function sustainably in their own right.

Alley-cropping (A-C) has been promoted as one such sustainable agricultural system (Kang & Wilson, 1984), among few, if any, feasible alternatives; it has been widely researched on the more fertile of tropical soils, often with emphasis upon nitrogen inputs (e.g. Kang *et al.*, 1990). However, it may be argued that the major unsolved problem of an alternative strategy to slash-and-burn/shifting cultivation lies in the widespread, leached, acid soils of the humid tropical forest zones (Jordan, 1989) (oxisols, ultisols, etc.) which together comprise some 63% of the the humid tropical zone worldwide and some 75% of the Central and South American Humid tropics (Sanchez, 1976). It can also be argued that, whereas findings emerging from studies on acid soils may well prove to be useful when translated to more favourable soil types, the reverse would not be true.

In the context of the history of humid tropical agriculture, the widespread colonisation and intense use of acid soils in the humid tropics is relatively recent. The colonisers are almost invariably poor and have commonly been displaced from the margins of other agricultural contexts; by failure of shifting agriculture, by warfare, by government policy or by economic necessity. Many indigenous examples of shifting agriculture, on the other hand, are thought to have arisen millennia ago on the fertile levées and terraces of tropical river basins where fishing and river transport would also have been integral to those cultures. The effects of population expansion alone would eventually have driven farmers away from the valley alluviums and on to the older, more deeply weathered and acid upland soils which are typical of the continental shields and uplifted landforms. Over extensive areas of the humid and sub-humid tropics, the classic effects of population pressure have transformed secondary forest/fallow mosaics into extensive perennial grassland, commonly as fire-climax vegetation. Such pressures of population are widely manifested in the cropping cycle as shortened periods under fallow and over-prolonged periods under cropping, or attempted cropping (Plates 1A & B).

Today, plantation agriculture on the fertile river terraces, among many other factors, has alienated many indigenous farmers from their traditional lands. Factors such as these, together with the more recent phenomena of transmigration schemes in Malaysia and Indonesia and examples of mass colonisation in the Amazon basin, have combined to place great pressure of subsistence agriculture on the acid soils of the humid and sub-humid tropics on a scale never seen before. Added to this, the aspirations of many former subsistence farmers have, not surprisingly, changed in recent decades to include the need for cash-cropping.

It is suggested here, that the role of NFTs in general, and *Inga* in particular, will have to be seen in the contexts of, firstly, acid-soil tropical rainforest ecology and, secondly, the ecology of the slash-and-burn clearance of those forests (Jordan, 1989).

The forest as a model of sustainablity

The adaptations to acid soils of the forest itself, one of the world's most productive ecosystems, is a starting point for the development of sustainable agroforestry systems. The tropical rain forest itself, is biologically a highly productive and sustainable ecosystem, yielding annually a dry-matter Net Primary Productivity (NPP) of perhaps 15–25 tonnes per ha (Kira cited in Whitmore, 1975). Many secondary forests, are scarcely less productive. This productivity contrasts with that of the newly burned swidden which might typically produce 2 tonnes of maize grain, as export, and a similar biomass of crop residue; moreover, this is a level of production which typically cannot be sustained on acid, leached tropical latosols beyond the second year. However, as all farmers and some researchers know, weed production following the first crop can be prolific.

The characteristics of acid tropical soils

The principal chemical characteristics of acid tropical soils have been well described by, among others, Sanchez (1976); Lal (1991) and Fisher & Juo (1995). This is the chemical context in which trees compete and have evolved tolerance (Plate 2A). Briefly summarised, these characteristics are:

i) High levels of exchangeable H, Al and Mn on the exchange complex; the toxicity of these cations to plants.
ii) Low Phosphorus (P) availability (expressed as the readily extractable orthophosphate anion).
iii) High P fixation (sorption) capacity of added phosphate (laboratory criteria).
iv) Low base-cation availability.
v) Low retention capacity (CEC) of those bases that may be added.
vi) High leaching rates of mobilised nutrients, whether derived from slash-and-burn, mineralisation or soluble fertiliser additions.

What has been described above is, to an acid-soil-intolerant plant, a fundamentally antagonistic soil-chemical environment and, moreover, one that actually competes with the roots for critical nutrients such as P. This is an environment characterised by low availabilty of major nutrients and low ability to retain and exchange those nutrients when they have to be added. It should, however, be emphasised that, in many forest soils, a marked stratification of surface soils has developed in which many of the more antagonistic properties of the mineral soil are strongly ameliorated by complexation with organic matter derived from litterfall and root turnover. The fine root concentrations that may be observed in these superficial layers may be interpreted as both a cause and a consequence of this stratification. The conservation of such superficial layers, possibly just a few cm in depth, remains one of the paramount conditions for a sustainable agricultural system in acid soils.

A number of marked characteristics have been observed in those rain forests growing on poorer tropical soils (Herrera *et al.*, 1978; Jordan, 1985; 1989), leading Jordan & Herrera (1981) to distinguish "oligotrophic" and "eutrophic" forest ecosystems; a distinction based mainly on soil base-status. The idea of phosphorus-limited ecosystems was carried further by Vitousek (1984) who, having compiled published accounts of N and P contents of forest litter, was able to demonstrate a strong correlation between litterfall and P content of litter in a substantial sub-set of tropical forest ecosystems.

Among the floristic adaptations to nutrient limitations listed by these authors are some (e.g. sclerophylly) which are probably of little use to a farmer, but a number of others appear highly significant; among them the following:

1) The P content of litter is lower than in more eutrophic forest systems (Vitousek, 1984). Nutrients are retained more tightly in the system and translocated prior to leaf-fall and, presumably, prior to fine root dieback, although information on the latter is lacking.

2) In the more extreme examples of oligotrophic systems (Herrera *et al.*, 1978) (e.g. "white-sand" vegetation), the tree rooting systems may tend to form superficial root mats; often clear of the mineral soil altogether and suffusing the decomposing litter where shading, moisture and physical protection allow this to happen. In less extreme examples, fine roots may be found clear of the soil surface under litter but, in any case, are frequently found concentrated in the uppermost few cms of soil.

3) Increased dependence of the forest trees upon symbioses with VA mycorrhizal fungi tends to correlate with increased oligotrophic status.

All of these features can be interpreted as tending towards more efficient retention, retrieval and recycling of P, in a soil context typified by low pH, low base-status, high H and Al toxicity (at least in subsoils), high P fixing capacity, shallow-rooting, etc.; they are part of a distinct, cohesive and global humid tropical pattern.

Well before the inception of the trials described here, phosphorus availability had been identified as one of the key constraints upon sustainable subsistence on these soils; firstly, because they typically possess low levels of available P; secondly, very high capacity for P fixation in the soil is common

(Sanchez & Uehara, 1980), especially in lower soil layers, and, thirdly, because these weathered and leached tropical latosols retain few primary minerals which, in younger soils, might have constituted a weatherable phosphorus resource.

Nearly all soil fertility research on alley-cropping systems concentrates on N cycling, yet in large areas of the tropics, it is more likely that phosphorus availability is the dominant factor limiting crop production. Jordan (1985, 1989) concluded from the San Carlos study that the post-burn decline in production he observed was most likely due to a reduction in P availability; and both Szott *et al.* (1990) and Palm *et al.* (1991) considered that P is the nutrient of most concern to the success of continuous crop production in A-C systems on ultisols.

As long ago as 1960, in their much quoted seminal volume, Nye & Greenland (1960) cited P availability as the most likely limiting factor in shifting cultivation in the humid tropics. Aspects of the phosphorus ecology of natural oligotrophic rainforests, slash-and-burn agriculture and alley-cropping are described in more detail in Hands *et al.* (1995) and, since the date of that paper, more data (see below) confirm the primacy of phosphorus in the sustainability of subsistence systems on acid soils of the Humid Tropics.

How trees adapt to acid soils

The roots of certain tree species tolerate or respond to acid soils in a number of ways, as described by Fisher & Juo (1995); trees are capable of resisting, complexing or sequestering Al^n ions. We can add "avoiding" to the list, for this is at least one outcome of root mat formation. Much more work on root physiology is needed before a clear picture emerges of the different strategies evolved by trees, but, from the characteristics given above, the following may be inferred:

1) The high dependence of many legumes on Vesicular-Arbuscular mycorrhizae (VAM) probably reflects their need for phosphorus in support of N fixation and would pre-adapt them for strong competitive ability on acid soils. It appears to be the case, but needing clarification, that the presence of some VA mycorrhizae can lead to the mobilisation of forms of organic phosphorus which are not otherwise available to plants (Dalal, 1982; Jayachandran *et al.*, 1992). High dependence on mycotrophism also implies a condition that systems be managed in the absence of tillage or soil compaction (which are known to disrupt the functioning of VAMs) and obviously, in the absence of fungicides. Mulching with organic matter would be expected to encourage and maintain mycorrhizal development (Dalal, 1982).

In the few studies that have been conducted on mycorrhizae and legume root systems, a number of *Inga* species were shown to have very high VAM infection densities (St John, 1980; Azcon, pers comm). In the absence of more specific data for individual species and soil types, the working assumption would be that *Inga*, as a highly dependent mycotrophic perennial, would tend to maintain high densities of mycelium and, presumably, spores. This could be expected to benefit any mycotrophic intercrop. Certainly, in the two trial sites outlined here, beans responded strongly to alley-cropping with *Inga*, and this may have been a factor.

2) That durable mulch material from certain legume tree species can encourage shallow or surface-rooting by physically protecting fine roots from desiccation and direct sunlight. Where rock phosphate and other supplements must be added to the cropping system, nutrients will be more efficiently retained and recycled by those raised fine roots (both of the trees and of the crop) than in bare soil or tillage systems in which the roots must compete with an antagonistic soil chemistry for available phosphate. It may therefore be the case that mulching for physical protection of the soil surface would constitute one of the overriding aims of a cropping system (such as alley-cropping), unless that protection is achieved, as it is under some perennial systems (e.g. coffee under shade), by a combination of shade and litterfall from the trees. In the case of alley-cropping with *Inga*, the triple aims of physical protection of the soil's structure, weed control and efficient nutrient retrieval are all achieved by the same physical factor; namely, durable mulch.

In view of the fact that *systems* research into the uses of NFTs generally, in acid soils, is at a very early stage, all that can be claimed is that the few case studies cited below provide the outlines of a far-from-complete picture. The spectra of species, techniques, constraints and possibilities that are available to farmers remain largely theoretical. Systems that are too finely tuned and which depend, for example, upon a particular selected provenance of NFT are unlikely to possess the resilience and adaptability for long-term success with farmers. Too little is known about the functioning of *whole systems* for any easy assumptions to be made about exactly what it is we are asking the trees to accomplish. The merits and demerits which have been claimed for alley-cropping are tabulated in Tables 2 and 3, together with some relevant arguments and qualifying statements:

THE ROLE OF *INGA* IN ALLEY-CROPPING AND THE RESTORATION OF DEGRADED SOILS

Almost all examples of alley-cropping (A-C) with NFTs on acid soils exist on experimental stations or as on-farm trials. The system is under investigation and development as a direct response to the need for sustainable systems in the context of failed slash-and-burn agriculture. However, early experiences with a number of *Inga* species planted in alley configuration have shown that they are also effective as a form of managed fallow or as a remedial measure for the regeneration of degraded cattle rangeland into cropping use.

Findings from Yurimaguas; lowland Costa Rica and elsewhere indicate that sustainable staple crop production in this acid-soil context will only be possible with the addition of critical nutrients. Soil supplements such as rock phosphate, dolomitic lime, etc. are likely to be the forms available, if at all, to resource-poor farmers and the crucial function of the trees in the cropping system, apart from N fixation, may be that their presence enables those supplementary nutrients to be retained and recycled tightly in the system and thus to be applied in small, affordable quantities.

Some acid-soil-tolerant NFTs respond strongly to light additions of these supplements and it may be reasonable to assume that systems in which they comprise a substantial component should be self-sufficient in nitrogen by

TABLE 2. Alley-cropping with *Inga*: Claims, comments and observations.

FACTORS OF A MAINLY PHYSICAL NATURE	COMMENTS
1. Permanent mulch cover from slowly decomposing prunings protects the vital surface soil layers from extremes of temperature and loss of its organic matter due to desiccation/saturation episodes (The "Birch* Effect").	In open soil, these factors tend rapidly to degrade the quality of SOM reserves and hence lead to the loss by leaching of the nutrients held in those reserves. The contention is supported by temperature observations.
2. Permanent mulch cover breaks the erosive force of heavy rain and preserves the structure and permeability of the surface soil aggregates.	Widely observed property of mulches in general. Supported by informal observations at the two sites outlined below.
3. Permanent mulch cover breaks the capillary pathways by which the soil loses moisture during drought periods; thus maintaining cool, consistently moist sub-mulch conditions and high levels of biological activity.	" " " "
4. Raised fine rooting systems: reproducing the patterns found in the rainforest. Physical protection afforded by deep mulch enables the system's fine roots to suffuse the organic surface soil layers and the mulch itself.	This has a number of very important physical, chemical and biological effects. See San Juan site data, below.
FACTORS OF A MAINLY CHEMICAL NATURE	
5. High organic matter inputs ameliorate, in the uppermost soil layers, the principal antagonistic soil characteristics associated with acidity.	SOM can complex aluminium ions; important in countering P fixation.
6. High inputs of fresh foliage would be expected to maintain high levels of the more active, labile forms of SOM.	Predictable, but not actually demonstrated.
7. Efficient retention and recycling of plant nutrients within the system.	Some evidence to support this; more needed.
8. Efficient retrieval of those plant nutrients that have to be added to the system.	P added as rock phosphate recovered in *Inga* prunings; San Juan data.
9. Nutrients raised from deeper soil layers.	Probably not true in the case of very superficial rooting patterns in acid soils.

* Birch (1960)

TABLE 2 continued.

FACTORS OF A MAINLY CHEMICAL NATURE (CONTINUED)	COMMENTS
10. Nitrogen inputs, derived from biological fixation, from the decomposition of fresh foliage.	Difficult to quantify, but the claim is consistent with observed high foliage production. Fine root and nodule dieback observed following pruning.
11. Much of the cation exchange capacity (CEC) in an acid soil resides in its soil organic matter (SOM) content and in its SOM/clay complexes. High OM inputs should maintain or enhance CEC.	Probable, but not proven in alley-cropping.
12. Carefully timed pruning claimed to present readily decomposable OM to the developing crop.	Probably not true in the case of recalcitrant *Inga* foliage.
FACTORS OF A MAINLY BIOLOGICAL NATURE	
13. Shallow and mulch rooting could reduce or nullify the effects of root pathogens.	See observations outlined for the two experimental sites below.
14. Increased biological diversity within the alley system compared to monocropping.	This is self-evidently true, but its significance is yet to be fully understood.
15. Weed germination and growth smothered by deep mulch.	One of the most important findings. See case histories, below.
16. The slight weeding that is necessary is easily carried out.	Consistently observed at both sites (see below). Weeds that do establish share the shallow rooting pattern of the trees and crops; they are easily hand pulled.
17. Extrafloral nectaries on *Inga* attract a wide diversity of insect predators; wasps, aggressive ants, etc.	Many species observed; some evidence of benefit to intercropped cultivars.
18. High levels of invertebrate activity beneath the mulch may enhance weed-seed predation.	Suggested by observations in alleys; not formally proved.
19. Maintenance of soil symbionts by the legume trees. (eg. VA mycorrhizae, rhizobia, etc).	High infection density of VA mycorrhizas in *Inga* roots.

TABLE 3. Alley-cropping: possible disadvantages; possible responses.

POSSIBLE DISADVANTAGE	POSSIBLE RESPONSE
1. Trees shade crops.	Very high potential for shading by *Inga*. Responses: prune shortly before sowing of the crop; reprune some weeks into crop development. Alley alignment and management have an effect here; also stem height is important. See below for options.
2. Trees can compete with crops below ground.	Significant fine root dieback observed at trial sites (see below) following pruning; presumably reinforced in second pruning. Root pruning found to affect yields at Yurimaguas but may not be necessary if 2 prunings carried out, as above.
3. Legume trees could harbour the same pathogens that affect other legumes; eg. beans.	Root knot nematode infection (*Meloidogyne spp.*) observed in beans in clear plots and in deeper roots in alleys; also observed in *Erythrina fusca* (see below). No infection observed in any *Inga* species under trial at either site in CR. Infected plants able also to root into mulch were little affected by the pathogen; this may be a very important, and unexpected effect of deep mulch. Nature of *Inga* resistance unknown.
4. The use of more durable mulch species could acidify the soil due to the nature of decomposition.	Little is known about long-term effects; presumably they are easily corrected by light additions of wood-ash, lime, etc. In the case of raised rooting systems, the soil volume in question is small and, hence, easily ameliorated.
5. Fast growing trees could acidify the rooting zone by very efficient base extraction.	Possible: but Ca, Mg, K, etc. returned in prunings; soil CEC enhanced by OM inputs. There could be a long-term system requirement for dolomitic lime. Long-term K economy unknown. Cash-crops could be needed to purchase light soil supplements as a minimum condition of sustainability.
6. A-C could involve more labour than open-plot systems.	A little more labour required in the year of establishment; nurturing small trees, etc. Once established, the key advantage lies in effective weed control under *Inga* mulch; very great labour saving compared to open plots.
7. A-C takes up more land area, compared to open plots, for a given area of crop.	True, but usually unimportant compared to the very high fallow/crop ratio of shifting agriculture or bush-fallow systems. No farmer in present trials (see below) has expressed any reservation in this regard; they are more concerned with yield per unit *effort* than per unit *area*.

63

biological fixation. At the very least, phosphorus and base cations removed in cropping will have to be replaced and it is useful to envisage successful alley-cropping systems on these soil types as efficiently retaining and recycling P, K, Ca, Mg, etc. rather than raising them from deeper soil layers which is commonly invoked as explanation for the functioning of alley-cropping on more favourable soils.

Although interest in the uses of *Inga* in alley-cropping appears to be growing, comparative data are somewhat scant. In particular, data describing the performance of *whole systems*, over long periods of time, are needed, together with data showing the long-term fates and fluxes of soil supplements which have been added to the mulch of a fully functioning alley system. Eventually, we could hope for a broader picture on how individual species or species groups are responding. The issue of 'sustainability or non-sustainability' could hinge on this knowledge. There appears to be little doubt that minimal soil supplements are an unavoidable condition for sustainability in any agricultural system. The key difference will be between maintenance supplements to a tight, "parsimonious" system and a complete fertiliser strategy. The latter is, in fact, an attempt at chemical domination of the soil and carries with it all the attendant costs and leakages which may be tolerable in other contexts and economies, but not in this.

Field trials in lowland Costa Rica

The case histories outlined below cover two long-term alley-cropping/clear-cropping trials in the humid tropical lowlands of Costa Rica. Principal NFT species in the alleys were *Erythrina fusca* and *Gliricidia sepium*; a choice of species which was based upon published knowledge which was current at the time of planning the experiments. Supplementary experimental plots were added as a means of gaining experience with *Inga* and, based on earlier observations, as a first-order screening of *Inga* types. In total some eight *Inga* species were planted and monitored in alley-cropping configuration, mainly at the San Juan site. Reference is also made below to a number of other trials, notably at Yurimaguas, Peruvian Amazonia.

The two Costa Rican sites are:

1) **Finca La Conquista** (La C), Sarapiqui. Costa Rica.

TOPOGRAPHY: 40 m. above sea level; uplifted former alluvial terrace; topographically flat and uniform. Rainfall infiltration and percolation rates are very rapid; no surface-flow (or erosion) was observed over the 3 years' trials.

RAINFALL; VEGETATION: 4,200 mm; Lowland wet tropical forest (Holdridge).

SOIL: Ultisol (pH in $CaCl_2$, 3.9). Very marked stratification of upper organic layers (0–2 cm; 2–10 cm; 10–20 cm, etc.) over orange/yellow mineral subsoil.

SITE HISTORY: Lowland wet tropical rainforest until logging in the 1950s; then remnant and secondary forest interspersed with subsistence agriculture and clearance for cattle. Secondary forest cover on the site was slashed-and-burned in approximately 1985, followed by sporadic cropping. Reslashed-and-burned by the Cambridge project 1989. Remnant trunks were reduced by chainsaw and removed by hand. Alley plots, clear plots and the first maize crop were established immediately after this burn. Based on observations of its performance locally, identical alleys of *Inga edulis* were established between two of the main experimental blocks (see Plates 2B–D).

2) **Co-operativa San Juan** (Co-operativa San Juan or San Juan). San Carlos. Costa Rica. (Northern plains bordering the San Juan river).

TOPOGRAPHY: Virtually identical to the La Conquista site; uniformly flat uplifted former alluvial terrace.

RAINFALL: Approximately 4,000 mm; no surface flow was observed during the entire 7-year duration of the trials.

SOIL: pH 4.1; slightly less deeply-weathered than La Conquista.

SITE HISTORY: Approximately 25-year-old secondary forest cover slashed and burned April/May 1989; all trunks reduced by chainsaw and removed by hand (Plate 1C). Experimental layout identical to La Conquista; later supplemented by alleys of 8 *Inga* species.

EXPERIMENTAL DESIGN (common to both sites): Large experimental plots (400 m^2), set out on flat terrain, and comprising four each of the following four treatments: Alley-cropping or Clear-cropping; both with and without applications of phosphorus (P) as rock phosphate. At La Conquista, 80 kg P per ha (40 + 40). At Co-operativa San Juan, 100 kg P per ha (in 3 applications 40 + 40 + 20, in the Decembers of 1989, 1990, 1991).

Associated Crops: Maize sown immediately post-burn (June 1989) and every subsequent June (Plate 1D); beans sown each mid-December (La Conquista; 3 years'; Co-operativa San Juan; 6 years' continuous monitoring).

Alley Alignment and Configuration: both sites: East–West alleys; 2.5 m wide; 0.4 m within-row spacing (10,000 tree stems per ha).

Tree species:

La Conquista: *Erythrina fusca* and *Gliricidia sepium* in alternating rows. *Inga edulis* in pure stand; identical configuration. (10,000 trees per ha).

San Juan: As above; 2.5 m. alleys *Erythrina fusca/Gliricidia sepium*, same configuration. 8 *Inga* species (*I. edulis; I. oerstediana; I. goldmanii; I. punctata* (added later); *I. spectabilis; I. densiflora; I. marginata; I. samanensis*); 4.0 m alley width; 0.5 m within-rows (5,000 tree stems per ha); alleys aligned East–West, arranged around main experiment. Identical soil and management (Plate 3A–C).

PRUNING REGIME AND MANAGEMENT (common to both sites). The trials were intended to investigate the cycling and ecology of critical nutrients, especially phosphorus. Very heavy potential weed growth was controlled by slashing 5–6 times per year, followed, at least twice per year, by the spraying of Paraquat to control regrowth immediately before crop sowing. This was carried out for experimental reasons, to remove weed competition as a factor in crop performance and in the hope of highlighting differences in nutrient ecology. In this respect, the experiments were not a strict replica of local farming practice. Deliberate year-by-year repetition of the cropping regime was maintained in order to impose a steadily increasing nutrient stress on the system.

Pruning Height: 1.5 m.

Pruning frequency: Determined by crop requirements; 1 major pruning, following weeding, 1 week before crop sowing; followed by one light pruning 5–6 weeks later into crop-growth. i.e. 2 major and 2 minor, prunings per year.

Timing: June; December.

Manner of pruning: Care was taken to minimise ripping damage; clean cuts were made with a very sharp machete or pruning shears. Approximately 5–10% of foliage was left on the stem at each pruning, mainly as short sprigs. No pruning cuts were made close to the main stem(s); thus leaving short spurs (10–20 cm) to facilitate regrowth. The aim was to ensure minimal cambium exposure or damage; regrowth often occurred from the node below the cut.

RESULTS

All plots started each cropping episode from the same weed-free condition. Without the paraquat, the clear plots would rapidly have been overwhelmed by weed-growth, whereas the alley plots maintained a degree of weed control for the first (2–3) years. The *Inga* alleys, once fully established, achieved a virtually complete weed-suppression.

Data available: Net Primary Production (NPP) of entire agro-ecosystems, including estimates of below-ground (b-g) production and estimates of weed components.

La Conquista: 3-years' data. San Juan: 6-years' data. Both sites, all above-ground (a-g) NPP in *Inga* alleys for comparison with the same components of the *E/G* system.

The distribution of fine roots within the systems

A number of corings were made with a saw-tooth corer (100 mm internal diameter) in order, firstly, to obtain a first estimation of the dieback which had been observed to occur in the finest tree roots and associated nodules following pruning; and, secondly, to compare vertical distribution patterns between soil layers and between species. A second set of corings was carried out at the La Conquista site shortly before abandonment and 5-months after pruning (Table 4).

1) *Grain Yields: Erythrina fusca/Gliricidia sepium (E/G) alleys versus clear plots; Inga alley plots.*

Because of the longer-term nature of the trials at Co-operativa San Juan, most of the data cited here derives from that site. The patterns at La Conquista were essentially similar, but all yields were at a lower level than those from the freshly-burned swidden at Co-operativa San Juan; differentials in biomass production between the *Inga edulis* and *Erythrina fusca/Gliricidia sepium* alleys were virtually identical between sites (see Table 5, below). All data from the main experiments are in preparation for publication elsewhere (Hands & Bayliss-Smith, in prep.); the main trends and findings are summarised here. Fig. 1 & 2 show the basic grain yields as a percentage of the first postburn crop.

PLATE. 1. **A** Slash-and-burn on marginal hillslope land in Central America; near Montaña del Carbon. Olancho. Northern Honduras. Young maize, upper right. Few trees of the original rain-forest cover remain; with most of the vegetation into the scrub or perennial grass cover resulting from two, or more, burning cycles. **B** The same hillside 18-months later; the consumptive process further-advanced. The erosion potential is obvious. **C** The San Juan site; 1-year post-burn. In the foreground is an experimental clear plot with a light weed-cover. The ground immediately behind our field assistant was slashed-and-burned, but never used; in this picture it holds one year's "weed" growth. **D** The first maize crop following the burn at the San Juan site; the crop has completely dominated both the weeds and the developing *Gliricidia sepium* (right).

67

PLATE. 2. **A** Soil profile in a road cutting near La Conquista; formed under tropical rain-forest and cleared a few years before this picture was taken. A virtually structureless acid, leached latosol of the utisol/oxisol type. Current estimates of tectonic uplift rates, and other evidence, indicate that this soil would have been laid down as alluvium in a fluvio-lacustrine environment, controlled by interglacial sea-levels, some 80–100 thousand years ago. Major tree-roots are seldom encountered at more than a few decimetres depth; the topsoil organic matter has largely disappeared. Landless subsistence farmers are increasingly attempting to settle marginal soils of this type which are widespread in the humid neotropics. **B** The La Conquista site. Seedlings of *Inga edulis* recently-established in 2.5 m alleys. **C** The same hedgerows of *I. edulis*, 3-months after transplanting; paraquat was used on the grasses. **D** The same hedgerows; 9-months after transplanting and undergoing their first pruning. The longest branches had reached 4.3 m and the canopy had completely closed.

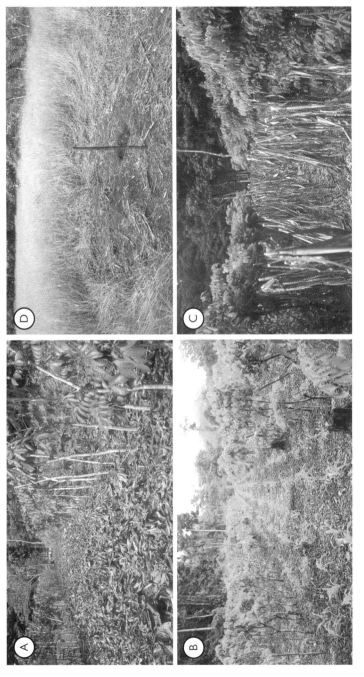

PLATE. 3. **A** San Juan site: 4-metre alleys of *I. densiflora* (foreground); *I. marginata* and *I. samanensis* (background). 16 months after transplanting and shortly after the first pruning. Mulch cover and weed-control are virtually complete; 5–8% of foliage left on the trees. Hedgerow height 1.5 m. **B** 4-metre alleys of *I. marginata*. Maize newly-emerged through the mulch, approximately 3 weeks after sowing and some 2–3 weeks before the second of the two prunings. Weed-control is virtually 100%. **C** The same alleys of *I. marginata*; the maize doubled-over at the point-of-harvest. Weed-growth is almost zero; no herbicides were used. The hedgerows have begun a 3-month period of regrowth before the next cropping cycle. **D** A weed-sampling quadrat at the San Juan site; May 1996. Experimental clear-cropping plots showing 2–3 months weed-growth, dominated by perennial grasses. Previous weed-control was by slashing, followed by a spraying of paraquat on the regrowth.

69

PLATE. 4. **A** The San Juan site; May 1996. 4-metre alleys of *I. edulis*, 2 months after the removal of the previous bean-crop. No herbicides were used on any *Inga* plot at this site. The sampling quadrats contained no weed-biomass to sample. **B** Mulch of *I. edulis* scraped aside (recent and older prunings) to show fine-roots and nodules clear of the mineral soil and suffusing the mulch. **C** Management of *Inga* hedgerows: *I. edulis* 3–4 weeks post-pruning, showing older unpruned foliage (above) and the fresh regrowth from clean-cut spurs (below).

San Juan site: Annual maize yields per plant

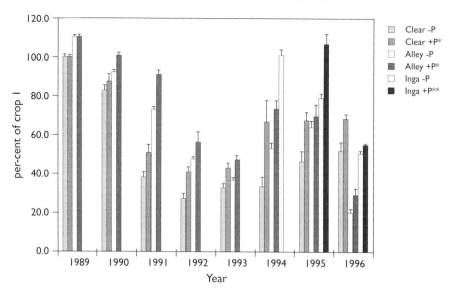

FIG. 1. The San Juan site: **Maize** grain production under clear-cropping and alley-cropping with various hedgerow species. Annual dry-weight grain yields per plant, expressed as a percentage of the first post-burn crop. The years 1994 and 1995 included very intense dry periods followed by unusually heavy rainfall during the wetter season. It is assumed that these contrasts have produced a 'Birch effect' (Birch, 1960) on yields.

E/G ALLEY AND CLEAR PLOTS; + AND − ROCK P.

Maize grain is taken to be the critical indicator; data refer to yield per plant (strictly speaking: per "sowing-point"; usual practice is to sow 2–3 seeds at each point):

Maize (Fig. 1)
  i)   Maize yields declined in all experimental plots from the level of the immediate post-burn crop.
  ii) Control (clear, -P) plots showed the classic post-burn decline in yields to a low steady-state in years 3–4.
  iii) Yields per maize plant responded strongly to rock P in both alley and clear treatments.
  iv) Yields per plant in the alley +P treatment exceeded those in the clear +P plots for the first 3 years; equalled or exceeded them in years 4, 5, 6 and 7; and were significantly lower in year 8.
  v)   Sub-plot experiments within the main plots showed no yield or biomass responses to lime + cation (Mg, K) additions in year 4 following the burn. The only responses in any plot have been to rock P.
  vi) The E/G alleys held 3 rows of maize at 0.75 m spacings. Depending on management, the two outer rows tended to be more or less suppressed by proximity to the tree-lines. Any yield differences were mainly provided by an enhanced performance in the centre row. The questions raised by this arc examined below under "Management Options".

71

Beans (Fig. 2)
  i) Bean yields were variable year-by-year, appearing to be strongly affected by seasonal factors (rainfall at a critical growth stage?). They showed positive responses to a-c in the first 5 years but dropped below their respective clear-plot yields in years 6 and 7.
  ii) As with maize, beans responded strongly to rock P applications and, for 5 years, most strongly to the alley +P treatment.

2) *Net Primary Production*
  i)   All ecosystem components, except *Gliricidia sepium* in the alley plots, responded strongly to rock P.
  ii) Weed biomass production proved to be potentially the largest system component, partially replaced by the trees in the alley plots.
  iii) Weed biomass showed a strong response to rock P.
  iv) NPP was sustained at a higher level, and for longer, in the *E/G* alley + P plots, but all systems showed a decline from year 2.
  v)   Prunings production in the *E/G* alleys was never sufficient to provide permanent mulch cover and declined from year 2. *Erythrina fusca* responded strongly to the rock P and came closest to providing complete cover. *Gliricidia sepium* did not respond to the rock P applications.
  vi) 5 years after the last application of rock P (Dec. 1991), most system components were still showing a residual P response.
Complete NPP data, over 6 years, are to be published separately (Hands & Bayliss-Smith, in prep.)

3) *Weeds*
  i)   Weed growth, in spite of a rigorous weeding regime, was the largest system component in the clear plots.
  ii) Weed growth, without rigorous slashing and spraying, would rapidly have overwhelmed the clear plots by the end of year 1.
  iii) If the cropping and pruning regime had been reduced to one episode per year, weed growth in the alleys would have been suppressed by a combination of mulch and shading.
  iv) Weed growth responded strongly to rock P additions.
  v)   Weeds in the clear plots were comprised of almost 100% grass species (*Paspalum sp.*; *Panicum sp.*); whereas weeds in the alleys were comprised of more herbaceous species (which are easier to control with a machete).
  vi) Weed-growth in the alleys was significantly suppressed by the more recalcitrant *Erythrina fusca* mulch during the first (2–3) years; thereafter increasing as mulch production declined in the *E/G* alleys.

Inga edulis alleys at the La Conquista site (Plates 2B–D)
  i)   At La Conquista, the 2.5 m alley width with *I. edulis* proved too narrow for maize.
  ii) The highest bean yields were in these alleys.
  iii) Beans in the *I. edulis* alleys showed no signs of root-knot nematode (*Meloidygone sp.*) infection, whereas bean roots in all other treatments were commonly infected. Root knots were present in *Erythrina fusca*, but were absent in *Gliricidia sepium* and *I. edulis.*

San Juan site: Annual bean yields per plant

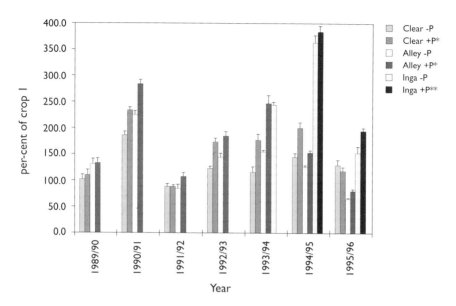

FIG. 2. The San Juan site: Bean grain production under clear-cropping and alley-cropping with various hedgerow species. Annual dry-weight grain yields per plant, expressed as a percentage of the first post-burn crop. Bean yields appear to highly variable year-to-year; probably due to seasonal factors such as rainfall or insolation at a critical time in the growth period. No long-term trend is discernible in these data, but yields have responded strongly to both alley-cropping, especially with *Inga*, and to rock phosphate. This is a very important and valuable basic grain crop in Central America.

iv) Weed biomass was virtually zero in the *I. edulis* alleys; the soil was permanently covered by mulch.

v) Those few weeds that were able to establish through the *I. edulis* mulch shared the same shallow-rooting characteristic as the trees and crops; because of this they were easily hand-pulled.

vi) Data for total biomass production in the tree components of the *E/G* and *I. edulis* alleys (excluding fine roots) showed great differences between the two alley types; NPP in the *Inga* for year 3 was approximately 25 tonnes per ha compared to 5.1 for the *E/G* alleys (data in Hands *et al.*, 1995).

Inga alleys at the San Juan site
Alleys 4 m wide; inter-tree spacing 0.5 m, 8 *Inga* species tested. Sub-plot treatments established in *I. edulis* alleys: with and without phosphate; 100 kg P per ha, as rock P.

i) Both maize and beans grown in *I. edulis* +P sub-plots maintained variable, but acceptable, grain yields for the 3 years which were monitored before the site was abandoned.

ii) *Inga edulis* foliage (leaf and stem prunings) responded to rock P applications both in biomass produced and in P content (Hands, unpublished data). Significant additional quantities of P were thus recycled through the mulch (see Table 6).

iii) Weed-control was virtually 100% in all *Inga* plots (see Plates 3B–D, 4A).

iv) Biomass production of prunings was maintained at a high and sustainable level over the 4 years of monitoring; some species were exceptionally productive (see data below) and all *Inga* species were significantly more productive than the *E/G* alleys.

v) As at the La Conquista site, beans responded particularly well to being grown in the *Inga* alleys.

vi) Informal observations indicate that a much higher biomass and diversity of invertebrates are to be found beneath the permanent mulch of the *Inga* alleys compared to the transient mulch cover of the *E/G* alleys. This could be an advantage or a disadvantage, but could be a factor in the mortality of weed seeds beneath the mulch.

vii) The 1991/92 bean sowing was virtually destroyed by slugs in the main *E/G* and clear plots. Beans growing in the only extant *Inga* alleys (*I. edulis*) were not attacked.

Fine root distribution and die back

In the *Erythrina fusca/Gliricidia sepium* alleys, corings taken before and after pruning showed that dieback resulting from pruning in the finest roots amounted to almost 100% in the uppermost layers (Hands, unpubl. data). As a first approximation, relating this input to foliage accumulated over the same period between prunings, gave a figure for below ground inputs as some 12% of leaf biomass. More data would obviously be desirable, but little information has hitherto been available, and serves as a best estimate for the time-being. As a high proportion of this input is comprised of nodules, it could represent an important source of readily-decomposed nutrients. Similar information relating to inputs in the *Inga* alleys is lacking. Corings taken at the La Conquista site shortly before the experiments were abandoned, yielded an interesting picture of the differences in fine root distribution between the *E/G* and *I. edulis* alleys under identical configuration and management (Table 4; Plate 4B)

DISCUSSION

These basic experiments took the form of a comparison between clear-cropping and alley-cropping within alternating rows of the legumes *Erythrina fusca* and *Gliricidia sepium*. The distinction between these two systems was somewhat artificial insofar as rigorous weed control was imposed upon all plots in order to allow differences in nutrient-supply to become expressed in crop yields. Because of very intensive weed-slashing, the comparison is actually that between two high OM-input systems; one based upon the slash-and-mulch of weed biomass; and the other based upon the prune-and-mulch of NFT foliage. In this sense, the hedgerow trees may be viewed as woody "weeds" under the farmer's control. Given the high levels of OM inputs in both systems, it is surprising that any yield differences were observed at all.

TABLE 4. Nitrogen-fixing Trees in Alley-cropping configuration. La Conquista. Sarapiqui. Costa Rica. (2.5 m alleys; 0.4 m within-row spacing (10,000 trees per ha); year 3). Distribution of fine root biomass between mulch and soil layers. Grams (dry-weight)/m². Species differences: *Inga edulis* and *Erythrina fusca/Gliricidia sepium.*

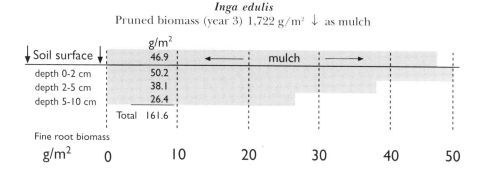

Inga edulis
Pruned biomass (year 3) 1,722 g/m² ↓ as mulch

↓ Soil surface ↓	g/m²					
	46.9	←	mulch	→		
depth 0-2 cm	50.2					
depth 2-5 cm	38.1					
depth 5-10 cm	26.4					
	Total 161.6					

Fine root biomass

g/m² 0 10 20 30 40 50

Erythrina fusca/Gliricidia sepium **(alternating rows)**
Pruned biomass (year 3) 404 g/m² ↓ as mulch

↓ Soil surface ↓	g/m²					
	0.0	←	no mulch	→		
depth 0-2 cm	27.9					
depth 2-5 cm	19.1					
depth 5-10 cm	9.6					
	Total 56.6					

Fine root biomass

g/m² 0 10 20 30 40 50

*Core-sampled Oct. 1992; 5 months' regrowth since pruning

Those differences that were actually observed seem to have been due to an enhanced P nutrition in the alley system (Hands *et al.*, 1995), coupled with the likelihood that the *E/G* foliage, and accompanying fine root dieback, contained more nitrogen.

The real-life situation is very different. Few, if any, farmers would sustain weed-slashing, without herbicides, in the open plots, beyond the end of the first year. Common practice, in the absence of fresh forest fallow in which to "shift", would be to leave the weeds and second-growth for 2 or 3 years, if space allowed; then to reslash-and-burn. This would probably provide a crop, albeit a reduced crop, to satisfy the urgent short-term need; but it is an essentially consumptive strategy which is seen widely to fail in the long-term. The *E/G* hedgerows, at 2.5 m alley width, were unable to maintain a mulch cover beyond a few weeks, whereas the *Inga* hedgerows, at 4.0 m width were able to maintain a permanent cover within 12–16 months of transplanting from the nursery.

Trials on farmers' fields; early impressions

Trials of *Inga* A-C with farmers, as a follow-up to the more scientific phase described here, are (1997) only at a developing stage, but farmers have seen a number of demonstrations in the experimental plots which are now mature and fully-functioning. The indigenous farmers with whom the trials (40) have begun, appear to regard a fairly wide spectrum of yields as acceptable, but the factor that does appear important is the virtual elimination of weeds in the *Inga* alleys. At demonstrations in the San Juan site, the health, appearance and pod-loading of the individual bean plants interested them greatly, but they were unconcerned with the fact that the space occupied by the tree lines would inevitably reduce yields *per hectare* of whole system. Compared to the problems of access to forest fallows, the inclusion of perhaps an additional 10–20% of land in order to accommodate the same number of maize plants in an alley system appears not to concern them. Yield-per-plant or, more precisely, yield-for-effort appears to be the important criterion with them. It does seem clear that, although questions of sustainability must inevitably hinge upon nutrient retention and cycling, the most immediate concern of farmers, and their proximal reason for shifting, lies in the intolerable burgeoning of weeds in their swiddens.

One unexpected comment came from some collaborating indigenous farmers in Honduras, to the effect that a-c might constitute a "mejora", under Honduran law, this is a legal instrument which could strengthen their claim to tenure-of-use of the land they occupy.

Crop pests

In the main San Juan trial plots, the 1991 (Dec.) bean crop was destroyed, shortly after germination, by slugs. However, beans sown in the *Inga* alleys were unaffected. The immediate solution was to deploy metaldehyde, protected within short lengths of bamboo, and to resow the beans. The key difference between the plots appeared to be that the slugs had been able to subsist throughout the previous year on weeds in both the open and *E/G* alley plots; whereas the *Inga* plots were both weed-free and slug-free. The slugs had been, if anything, favoured by the temporary protection of the sporadic *Erythrina* mulch during the occasional dry spells which are normally experienced earlier in the year during the period February to May.

The importance of surface-rooting in Inga

The paramount importance of surface-rooting to *Inga* was illustrated by two occurrences in the La Conquista site. As part of an investigation into VA mycorrhizal spore density and distribution, surface soil was removed to a depth of 5 cm only, and for a distance of 5 m along one side only, of one of the *I. edulis* hedgerows. The result was a severe (75%) reduction in foliage production, during the ensuing weeks, of all the 10 tree stems affected, and the deaths of 3 of those trees. This occurred even though the trees were able to continue normally on the unsampled side. A later soil-sampling alongside another section of hedgerow produced an identical suppression of foliage regrowth.

Care in pruning

Reports from CATIE (Sanchez, pers comm.) and elsewhere (Cochabamba, Bolivia), suggested that *Inga* species are unsuitable for use in a-c as they cannot withstand coppicing. This was not found to be the case in these trials, provided that care is taken to minimise the impact of the pruning and that no attempt is made to remove 100% of the foliage at any one time (Plate 4C). As far as is known, the energy-drain on an individual *Inga* tree can be high; not only does the tree have to support its own metabolism and regrowth, but also, the symbiotic load of the nodules, VA mycorrhizae and nectar production. It was found not to be necessary to strip the tree completely.

Connected with this, an investigation in the *E/G* alleys into fine root densities and distribution in surface-soil layers, revealed a high mortality, approaching 100%, of the finest roots (< 1 mm approximately) following pruning. Slightly coarser roots (> 2 mm diam. approx.), rising from the main laterals at approx. 10 cm depth and which appear to be conducting tissue, survived intact. It may be that the energy-drain on the tree following pruning is too great to support the VAM and that the only way of reducing that drain, once the fine roots are VAM-infected, is by dieback. Without exhaustive experiments, the measure adopted during these projects, i.e. the leaving of 5–10% of foliage at each pruning, appeared to work.

Sustainability in pruned biomass production

Trees grown in the stressful, and relatively exposed, conditions of the alley system are likely to be more subject to variations in climate than those in the protected and more-constant forest environment; also, the timing and frequency of pruning will affect annual NPP (see below under "Management Options"). The NPP data given here (Table 5) result from a somewhat intense, but consistent, pruning regime; and, although it would be desirable to continue the monitoring over as long a period as possible, these pruning data from 7 *Inga* species, over 4 years, shows no apparent diminution in annual productivity and far exceed those of the *Erythrina* and *Gliricidia* from the main plots. This, at least, is one indicator of sustainability

Alleys of *I. punctata* were established later than the others at San Juan and data derived from them are thus not strictly comparable, but this species required a long period for establishment at the San Juan site. It may be that provenance is important with *I. punctata*, as other trials (not reported here) in Honduras are showing that seedlings grown from riparian sources (richer soils) are failing to establish well in more-acid upland soils nearby.

Ability to export Phosphorus in the form of grain

In soil conditions and ecosystems which are P-limited, one possibly useful criterion for the evaluation of any agricultural system is its ability to export that limiting nutrient as grain. Similarly, efficiency of retrieval and recycling could be a useful criterion for forestry sytems which do not export P in such an immediate way. A long series of P content and biomass determinations on material from the San Juan site is still underway (1997), and will be the subject of a separate publication (Hands, in prep.). However, some provisional data are included here which lend support to these ideas (Tables 6 and 7).

TABLE 5. Alley-cropping; La Conquista and San Juan sites. Lowland Humid Tropics, Costa Rica. Accumulated biomass of prunings (Branch + foliage) over 4 years: g (DW) per metre of hedgerow. Growth periods from transplanting date. * Mean value of two species in alternating rows. s.e. = standard error.

SPECIES	CONFIGURATION	SITE	g (dry weight) of prunings (branch + foliage) per m of hedgerow									
			12 months' growth	s.e.	20 months' growth	s.e.	24 months' growth	s.e.	36 months' growth	s.e.	48 months' growth	s.e.
I. edulis	4.0 m alley width × 0.5 m (5,000 trees/ha)	San Juan	2,290	88	6,773	381	7,869	386	11,108	430	14,632	462
I. oerstediana	"	"	4,486	750			10,742	938	16,012	967	21,192	995
I. goldmanii	"	"	2,074	152			4,151	265	6,950	309	9,060	332
I. spectabilis	"	"	1,690	105			3,973	242	7,054	336	11,534	376
I. densiflora	"	"	728	84			3,702	247	9,878	485	13,344	524
I. marginata	"	"	3,520	295			6,158	400	10,188	515	13,572	532
I. samanensis	"	"	2,507	127			4,864	325	8,304	395	11,576	432
Erythrina fusca	2.5 m alley width × 0.4 m (10,000 trees/ha)	"	1,190	42	2,370	54	2,735	63	4,022	74	4,915	81
Gliricidia sepium	"	"	770	25	1,934	41	2,720	60	4,837	76	6,290	88
Erythrina/Gliricidia*	"	La Conquista	n.a.		2,150	75	n.a.		3,436	72		
I. edulis	"	"	n.a.		5,925	260	n.a.		10,275	356		

TABLE 6. San Juan site 1995; Phosphorus (P) returned to the soil as prunings or slashed material: Selected *Inga* alleys compared to clear control plots; with and without added rock phosphate. Kg/ha/yr.

TREATMENT		MATERIAL	Kg P/ha/yr	s.e.
			Returned to the soil 1995 ↓	
Control plots	-P	Weed a-g biomass *	3.32	(0.12)
Control plots**	+P	" "	7.97	(0.33)
Inga alleys; 4.0 m				
I. edulis	-P	Leaf + branch prunings	27.66	(1.50)
I. edulis***	+P	" "	37.30	(0.77)
I. oerstediana	-P	" "	41.43	(1.56)
I. marginata	-P	" "	27.97	(1.72)

*above-ground biomass. ** 100 kg P ha⁻¹ added as rock phosphate. *** 100 kg P ha⁻¹ added; Dec. 1994, as rock phosphate.

TABLE 7. Alley-cropping with *Inga* versus Clear-cropping in P-deficient soils; with and without additions of rock phosphate. The ability of the systems to export phosphorus in the form of grain.

San Juan site; 1995; Kg P/ha/year

Plot Type	Control (clear-cropping)		*Inga edulis* Alleys (4 m width)	
Treatment: + or - rock P →	-P	+P	-P	+P
Plants/ha				
Maize*	21,333	21,333	13,333	13,333
Beans*	60,000	60,000	56,250	56,250
P exported in grain; Kg/ha:				
Maize	3.30 (0.80)	6.62 (0.78)	3.85 (0.22)	6.54 (0.73)
Beans	2.42 (0.26)	3.13 (0.34)	4.55 (0.32)	6.18 (0.37)
Total	5.72 (0.84) ↓	9.75 (0.85) ↓	8.40 (0.39)	12.72 (0.82)
Total P exported in grain per unit of cropped area alone: Kg/ha	5.72 (0.84)	9.75 (0.85)	11.01 (0.49)	17.06 (1.23)

Standard errors in parentheses.

* Spacing within and between rows of crop plants: Maize: 0.75 m; Beans: 0.4 m. In the alley system, the 4.0 m alleys contain 5 rows of maize (sown in June) or 9 rows of beans (December).

In leached, P-deficient soils, both grain yield and Phosphorus content of the grain vary according to treatment. The ability of the system to export P in the form of grain is thus a more useful index of performance than yield alone. In this example (San Juan site; 1995), the phosphorus economy of the *Inga* alley system is clearly superior to that of clear-cropping, in spite of the fact that the latter is recycling significant quantitites of P through the weed component.

In the alley plots, one row of crop is substituted by one row of trees. Judging by the biomass production of the *Inga*, this, if anything, would appear strongly to bias the data in favour of the control plots, as it is likely that the crop rows adjoining the hedgerow will experience higher levels of below-ground competition in addition to shading. On the basis of cropped area alone, the *Inga* system is exporting significantly more P as grain than either control treatments (Table 7). In all the *Inga* alleys, the amounts of P returned to the soil as mulch (Table 6), and released by decomposition during the 3-month growth period of the maize, are commensurate with the amounts exported in a normally acceptable maize crop; this ignores the possibly significant amounts returned in fine root dieback. Whether such P is truly available to the crop is, of course, not proved by the data; but, at least, by this criterion, the *Inga* alley system possesses the fundamentals of sustainability.

THE DESIGN AND MANAGEMENT OF ALLEY CROPPING SYSTEMS USING *INGA*

System variables within the control of the farmer
1) Choice of species: As argued above, more-durable mulch species are desirable; but mixtures are feasible and might include a proportion of readily decomposable foliage species such as *Gliricidia sepium*, which is very well-known throughout Central and South America. The research project reported here has trials of pepper (*Piper nigrum*) grown on living supports of this species, between hedgerows of *Inga*. The pepper receives no agro-chemical inputs whatsoever and, in a weed-free environment, appears to be thriving.

The first choice *Inga* is likely to be the commonest local provenance; and commonly, the *Inga* species grown for shade over coffee or cacao. However, *I. edulis* is not endemic to Central America, yet appears to have undergone a widespread and spontaneous local adoption as an easily-established shade species.

There exists a wide spectrum of choice within which to choose, and alley-cropping systems with *Inga* appear to be rather forgiving, provided they are pruned with care.

2) Within-row density: Although more detailed work is needed, the indications are that, above a certain planting density, foliage production in an alley hedgerow tends to stabilise at a certain level per metre of hedgerow, while decreasing per tree stem. A useful spacing in the La Conquista and San Juan sites appeared to be about 0.5 m, which allows for some sporadic mortality without compromising mulch production; however, this was a somewhat intuitive choice.

3) Alley width: A balance needs to be struck between a wider spacing to minimise competition between the trees and crops and a narrower spacing to maximise weed-control; the working assumption being that the latter is the dominant factor. A general rule-of-thumb would then be: "As wide as is compatible with sufficient mulch production for permanent cover". Any over-dominance by the tree component can be controlled by varying other management practices, such as pruning height or frequency.

The 4.0 m alley width for the *Inga* at the San Juan site was also an intuitive choice based upon experiences at La Conquista. The most productive species of the San Juan trials (*I. edulis*; *I. oerstediana*) should be able to achieve permanent mulch cover at 5 m alley width.

4) Alley alignment: It can be argued that, in equatorial regions, an ideal alley alignment would be East-West, because this is likely to impose more competition for sunlight within tree rows than between the trees and the crops which will be in sunlight for most of the day. This could be important for maize, but beans at both the La Conquista and San Juan sites appeared little affected by a degree of shading, as the sun, in the Northern winter, swung to the southward of the hedgerows.

However, the need for a contouring alignment to counter the risk of erosion on slopes may override this ideal; and this factor, in turn, implies that greater care needs to be taken to reduce shading. It seems, therefore that decisions of this nature need to be integrated with other management variables; such as: "If there is a risk of shading from contoured tree lines, then reduce stem-height or increase pruning frequency".

5) Hedgerow height: The higher the stem, the greater and more rapid will be the recovery of foliage; and, probably, the more resilient the tree; whereas, the lower the stem is pruned, the fewer the available nodes for regrowth. *Inga* appears to require a moderate-to-high stem. Pruning-height and frequency are probably the easiest ways of altering the dominance or "presence" of the tree component in an alley system. The height options thus fall within the range: a) to favour the trees and mulch production, as high as is feasible (say up to 1.75 m); b) to disfavour the trees, coppiced low to the ground (but the suggested minimum with *Inga* would be about 1 m).

6) Pruning regime (Table 8): In practice, the timing and frequency of pruning are likely to reflect cropping needs, rather than any agenda set by the trees themselves; but there are some exceptions. For example, if experience shows that an alley system may only sustain one crop per annum, then tree growth in the period between crops may, in the case of some *Inga* species, be very vigorous, leaving few branches and little foliage on the lower stem. This, in turn, may leave the stem bereft of foliage when pruned back to the working height. (*I. edulis* at the La Conquista site grew to over 4 m in 9 months from transplanting). In this instance, it is advisable to carry out the pruning in two phases. It may be preferable to cut out the leading central stem, reducing shading and allowing the lower stem to resprout before pruning back the side branches. However, this may not permit enough light to strike low enough on the stem for this to happen. Alternatively, it may be better to cut out all the side branches over the final pruning height, thus leaving the central leader to maintain the tree whilst the lower stem resprouts. When this is clearly under way, the leader (called a "chimenea" in Costa Rican cafetales) can be taken out. The aim would be to time the second pruning for a week or two before the planned crop-sowing; this, in turn, implies a first-phase pruning some month to six weeks earlier. All this would have to harmonise with local

TABLE 8. Pruning frequency in alley-cropping; a spectrum of effects.

Frequency of pruning	
Low	**High**
Greater dominance of the system by the tree component	Lesser role of the tree component
Greater shading of weeds Less smothering of weeds by mulch	Lesser shading of weed More smothering of weeds by mulch
The greater overall production of pruned biomass	Less overall production of pruned biomass
Greater proportion of woody biomass in the prunings	Greater proportion of leaf biomass in the prunings
Greater likelihood that 2-phase pruning will be necessary	Little likelihood that a 2-phase pruning will be needed
The possibility that the system will produce firewood	The probability that the system will not produce firewood

perceptions as to what are, or are not, good pruning and sowing times. In any case, a light pruning will be needed to reduce competition perhaps some 4–6 weeks into crop growth.

As outlined above, manner of pruning is very important to the survival of the trees and any attempt at a 100% foliage removal will involve the risk of killing the tree. Similarly, the damage and ripping associated with pruning too close to the main stem must be avoided. Clean cuts with sharp tools, together with the leaving of short spurs, with some foliage, minimises this risk; also, the branches of some *Inga* species (e.g. *I. marginata*; *I. samanensis*) are characteristically more slender and appear to suffer much less setback as a result of pruning. It is hoped that more detailed knowledge of the pruning-tolerance of the species groups may be gained in the future.

7) Soil supplements

Rock Phosphate. As argued above and elsewhere (Palm *et al.*, 1991; Hands *et al.*, 1995), the minimum condition for any low-input, sustainable agricultural system, as a stable, alternative subsistence strategy to shifting cultivation, will be that maintenance supplies of phosphorus will have to be made cheaply available. For a number of reasons, the obvious source of this phosphorus input is rock phosphate; and it is suggested that this is more efficiently applied to the mulch in an alley system, rather than to the soil itself. At a national or regional scale, the logistical difficulties of this are clearly very great and involve social, political and economic issues which go far beyond the scope of this text; the condition itself is, however, ecological in nature (i.e. pertaining to plant ecology) and non-negotiable.

Lime. In addition to rock phosphate, one further long-term ecological condition which may have to be fulfilled on an acid soil relates to slash-and-burn agriculture itself.

One aspect of the short-term success of a slash-and-burn operation in a rain-forest swidden lies in the effect of the ash upon the accumulated organic reserves of the soil (SOM). In short, SOM that may be turning over very slowly may undergo an accelerated decomposition due to a temporary change in the pH of the immediate surface soil (Hands, 1988; Hands *et al.*, 1995). It is the release of N and P associated with these reserves that may be the key process in swidden agriculture. In a green mulch system such as A-C with *Inga*, it is likely that, over time, the soil will accumulate a wide range of SOM types with many differing decomposition characteristics. It could prove to be necessary for well-sustained maize yields, for example, that this pH-effect of the ash will have to be simulated, not by burning, but by lime or dolomitic lime (for the magnesium); and, possibly, by some source of potassium. The expectation is that A-C will retain and recycle these supplements better than any bare-soil alternative.

There is wide scope for experiment with all these options and experiences with *Inga* in alley-cropping indicate that these systems are rather flexible, resilient and forgiving. Supplements, as described above, are, of course necessary for long-term sustainability; but the system does not appear to collapse if they are witheld for a year.

THE USES OF *INGA* IN THE RECLAMATION OF DEGRADED SOILS; ABANDONED PASTURE; ETC.

The need to rehabilitate soils that were formerly under forest, and that have since degraded as a result of exposure and use as cattle rangeland, appears poised to become one of the priorities of the future. In this case, the hope would be that nitrogen-fixing tree species, whether planted in alleys or at "nurse tree" spacing for reforestation, can begin the process by which the soil's condition and OM-content can be returned to a state more closely resembling that of the original forest. *Inga* are light-demanding forest gap species (Pennington, this volume) and, although not pioneer colonisers of open-ground, they can function as such, once they are established as seedlings; some species do this very effectively. However, the system cannot fulfill the functions that are ascribed to it here unless its roots can encounter nutrients for retrieval and recycling. It thus appears very likely that the use of *Inga* in this remedial role, and with cropping as a desired goal, will require the help of soil supplements; a conclusion also reached, in the more general context of managed fallows, by Sanchez (1987) and Palm *et al.* (1991).

The project of which the Co-operativa San Juan and La Conquista trials comprised an early phase has begun a trial on a soil which has been exposed for 12 years since burning and which now appears to consist of bare, red sub-soil. The initiative for this came from the farmer himself who considers that he now has no other option than to give alley-cropping a trial. Given the long fallows which would be considered normal in indigenous shifting agriculture, the *Inga* alleys will face a protracted struggle, even to establish themselves, and leaving aside, for the moment, thoughts of crop production.

Also, a number of A-C experiments, reported in the literature, on open, but much less degraded soil contexts, have been maintained for some years before showing any yield response.

In this extreme, but nevertheless increasingly widespread, example, the addition of cheap sources of P, Ca, etc., could be enough to stimulate foliage production to a level capable of providing adequate mulch; which, in turn, would raise the fine root system; which, in turn, would use the supplements with increasing efficiency. Eventually, gradually increasing influence of the mulch should begin the replenishment of the soil's OM-content and quality to the state in which it had originally been 12 years before. This is the theory behind the trial.

The species in question here are *I. oerstediana* and *I. marginata*, in alternating rows; and both of local provenance. Both are readily available as seedlings taken from beneath the parent trees for transplanting into forestry bags, and later into the hedgerows. In common with the experiences outlined above at the San Juan site, and also elsewhere (Pennington, this volume), *I. oerstediana* has proved itself a very aggressive pioneer of open ground and seemingly outstandingly tolerant of soil acidity. *I. marginata* appears to need a little more help in early weed control, but once established, also appears excellent as a less aggressive A-C species; its root nodules are larger than those of *I. edulis* or *I. oerstediana*; moreover, its finer branching habit renders it more tolerant of pruning than most *Inga* species under trial at San Juan (see also Neill & Revelo, this volume, on the use of *Inga* for the control of old pasture).

ACKNOWLEDGEMENTS

None of the long-term work described here would have been possible without the support of the Commission of the European Communities under, initially, the DG XII program, Science and Technology for Development (STD); and, in latter stages, under the The Tropical Forests budgetary line of DG I. We are grateful to Dr Alfredo Alvarado and Ing. Gabriela Soto for collaboration with the University of Costa Rica and for the use of UCR laboratory facilities. The author is personally indebted to the many colleagues and staff of the Geography Department at Cambridge for help and encouragement over a number of years; and, especially to Tim Bayliss-Smith for help in innumerable ways. Many assistants and volunteers have been involved over the years with establishing hedgerows and with the laborious business of monitoring the output of many large experimental plots; special thanks go to Edward Coode, Nigel Ede, Kate Hands, Ben Hands, Simon Hetherington, Anna Lawrence, Helen Leggett, Noreen Matheson, Hannah Nuttley, James Sowerby and Jonathan Wibberley.

REFERENCES

Birch, H.F. 1960. Soil drying and soil fertility. Trop. Agric. (Trinidad) 37: 3–10.
Brookfield, H.C. 1988. The new Great Age of Clearance and beyond. In: J.S. Denslow & C. Padoch (eds.), People of the Tropical Rain Forest, University of Berkeley Press, 205–224.

Dalal, R.C. 1982. Effect of plant growth and addition of plant residues on the phosphatase activity in the soil. Pl. & Soil 66: 265–269.

Denslow, J.S. & Padoch, C. (eds). 1988. People of the Tropical Rain Forest. Smithsonian Institution / Univ. of California Press. Berkeley.

Fisher, R. & Juo, A. S. R. 1995. In: D.O. Evans & L.T. Szott (eds.), Nitrogen Fixing Trees for Acid Soils. Nitrogen Fixing Tree Res. Rep. (Special issue).

Hands, M.R. 1988. The Ecology of Shifting Cultivation. Unpubl. M.Sc. thesis, University of Cambridge.

Hands, M.R., Bayliss-Smith, T.P. & Bache, B. 1993. Experimental alley cropping systems in lowland tropical rain forest sites in Costa Rica: I; Biomass Production. Report to the Commission of the European Communities DGXII, Science and Technology for Development, Brussels. Department of Geography. University of Cambridge. England.

Hands, M.R., Harrison, A.F. & Bayliss-Smith, T.P. 1995. Phosphorus dynamics in slash-and-burn and alley-cropping systems on ultisols in the humid tropics: Options for management. In: H. Tiessen (ed.), SCOPE Final workshop proceedings. Phosphorus dynamics in Terrestrial and Aquatic ecosystems: A Global Perspective. Wiley.

Hecht, S.B. 1985. Environment, Development and Politics: Capital accumulation and the livestock sector in Eastern Amazonia. World Development 13: 663–684.

Herrera, R., Jordan, C.F., Klinge, H. & Medina, E. 1978. Amazon ecosystems: Their structure and functioning with particular emphasis on nutrients. Interciencia (Caracas) 3: 223–232.

Jayachandran, K., Schwab, A.P. & Hetrick, B.A.D. 1992. Mineralisation of organic phosphorus by vesicular-arbuscular mycorrhizal fungi. Soil Biol. Biochem. 24: 897–903

Jordan, C.F. 1985. Nutrient cycling in Tropical Forest Ecosystems. John Wiley.

Jordan, C F. (ed.). 1989. An Amazon rainforest: the structure and function of a nutrient stressed ecosystem and the impact of slash-and-burn agriculture. MAB/UNESCO Series 2, Parthenon Press.

Jordan, C.F. & Herrera, R. 1981. Tropical Rain-forests: Are nutrients really critical? Amer. Naturalist 117: 167–180.

Kang, B.T. & Wilson, G. F. 1984. Alley cropping: A stable alternative to shifting cultivation. IITA, Ibadan, Nigeria.

Kang, B.T., Reynolds, L. & Atta-Krah, A.N. 1990. Alley Farming. Advances Agron. 43: 315–359.

Lal, R. 1991. Myths and scientific realities of agroforestry as a strategy for sustainable management for soil in the Tropics. Advances Soil Sci. 15: 91–137.

Nye, P. & Greenland, D. 1960. The soil under shifting cultivation. Techn. Bull. 51. Commonw. Bur. Soils. CAB. Harpenden.

Palm, C.A., McKerrow, A.J., Glasner, K.M. & Szott, L.T. 1991. Agroforestry systems in lowland tropics: is phosphorus important ? In: H. Tiessen, D. Lopez-Hernandez & I.H. Salcedo (eds.), Phosphorus cycles in terrestrial and Aquatic ecosystems. Regional Workshop 3: South and Central America. SCOPE Workshop Maracay, Venezuela 1989. Saskatchewan Institute of Pedology, Canada.

Sanchez, P. 1976. Properties and management of soil in the tropics. Wiley. New York.

Sanchez, P. 1987. Soil productivity and sustainability in agroforestry systems. In: H.A. Steppler & P.K.R. Nair (eds.), Agroforestry: A decade of Development. ICRAF. Nairobi. Kenya.

Sanchez, P.A. & Uehara, G. 1980. Management considerations for acid soils with high P-fixation capacity. In: F.E. Khasawneh, E.C. Sample & F.J. Kamprath (eds.), The Role of Phosphorus in Agriculture. American Soil Association, U.S.A.

Schmink, M. & Wood, C.H. 1984. Frontier Expansion in Amazonia. Univ. of Florida Press. Gainsville.

St. John, T.V. 1980. Una lista da especies de plantas tropicais Brasileras naturalmente infectados con micorriza vesicular-arbuscular. Acta Amazon. 10 (1): 229 et sec.

Szott, L.T.; Palm, C.A. & Sanchez, P.A. 1990. Agroforestry in acid soils of the humid tropics. Advances Agron. 45: 275–301.

Vitousek, P. 1984. Litterfall, nutrient cycling and nutrient limitation in tropical forests. Ecology 65: 285–298.

Whitmore, T.C. 1975. Tropical Rain Forests of the Far East. Clarendon Press, Oxford.

CHAPTER 6. THE POTENTIAL OF *INGA* SPECIES FOR IMPROVED WOODY FALLOWS AND MULTISTRATA AGROFORESTS IN THE PERUVIAN AMAZON BASIN

JULIO C. ALEGRE, JOHN C. WEBER & DALE E. BANDY

INTRODUCTION

Shifting cultivation is the most common agricultural land-use system in the humid tropics, and is based on the indirect recycling of nutrients from woody biomass to cultivated crops (Nye & Greenland, 1960; Sanchez, 1976). In the Peruvian Amazon Basin, the upland soils are generally acidic and lack sufficient essential nutrients for sustainable, repeated crop harvests without nutrient inputs. Farmers deal with this situation by slashing and burning primary and secondary forests to release accumulated nutrients in the woody biomass. Decomposition of above- and below-ground biomass and nutrients in the ash typically allow a few years of cropping, and then the fields are either managed as woody fallows for several years before the next cropping cycle, or as sequentially enriched multistrata agroforests for production of timber and non-timber wood products (Denevan & Padoch, 1990). Most trees in these fallows and multistrata agroforests develop from natural regeneration of existing species in nearby secondary forests. Many farmers in the Peruvian Amazon Basin select and plant *Inga edulis* Mart. in their fallows and multistrata agroforests because they recognize its value for soil improvement, and it provides additional products and services (Potters, 1997; Brodie *et al.*, 1997).

The traditional slash-and-burn agricultural system is not sustainable for many farmers in the Peruvian Amazon Basin. In the rapidly growing Pucallpa region, for example, an average farm is about 22 ha in total area, with 2 ha/year under cultivation, 4 ha in different stages of managed secondary forest, 5 ha in managed primary forest, 10 ha in pastures, and the remainder in home gardens and boundary plantings (Sotelo-Montes & Weber, 1997). A typical production cycle starts with the slash-and-burn of a 2 ha parcel of a 5–20 year-old secondary-forest fallow. Several crops are cultivated for 1–3 years: mainly rice (*Oryza sativa*), maize (*Zea mays*), beans (*Phaseolus* spp.), cowpea (*Vigna unguiculata*), and cassava (*Manihot esculenta* Crantz), the number of years depending on the soil fertility replenishment achieved by the fallow and slash-and-burn, weedy species' characteristics and the invasion dynamics. Then the parcel is sown with plantain (*Musa* spp.) and selected fruit and timber trees, or managed as another secondary-forest fallow. The farmers' current supply of secondary forest will not be able to satisfy their continuous demand over time, and consequently they will be forced to (1) reduce the fallow period leading to greater soil degradation and lower crop productivity, (2) slash-and-burn their remaining primary forest, or (3) search for new forested land to slash and burn. This scenario of deforestation related to unsustainable slash-and-burn agriculture is seen throughout much of the humid tropics (Alegre *et al.,* 1986; ICRAF, 1995).

The objective of this chapter is to illustrate the potential of *Inga* species for use in improved woody fallows and multistrata agroforestry systems in the Amazon Basin, drawing upon research experience in Peru. Farmers in the Peruvian Amazon Basin use several species of *Inga* (Pennington, 1997). The International Centre for Research in Agroforestry (ICRAF) investigated farmers' preferences for agroforestry tree species, and identified *I. edulis* Mart. as a high priority species for agroforestry research and development in the Peruvian Amazon Basin (Sotelo-Montes & Weber, 1997). We focus on *I. edulis* in the Peruvian Amazon Basin throughout most of this chapter because it is a priority agroforestry tree species, and there is considerable research experience with the species in Peru. The chapter includes four sections: functions of improved woody fallows and multistrata agroforests, farmers' uses of *I. edulis* in these systems, research on *I. edulis* for these systems, and selection of *Inga* species for these systems.

FUNCTIONS OF IMPROVED WOODY FALLOWS AND MULTISTRATA AGROFORESTS

Improved woody fallows and multistrata systems have different functions in the ecological succession towards "climax agroforests" at the landscape level (Leakey, 1996). Improved woody fallows are examples of early-successional components of this "climax agroforest", while multistrata systems may be early-, mid- or late-successional components depending on their species composition and structure.

Improved woody fallows can have several functions: reduce the length of the fallow period while maintaining or increasing crop yields; reduce weed invasion; reduce soil erosion, maintain or increase soil fertility, and improve soil physical properties; and provide additional tree products for local use or sale. Farmers in the Peruvian Amazon Basin decide to fallow their fields and slash/burn another forest parcel when declining crop yields and increasing weed pressure no longer justify their labour investment (Denevan & Padoch, 1990). The slash-and-burn of a 20 year-old natural secondary-forest fallow is generally necessary to replenish soil fertility for a 3-year cropping cycle. Due to limited supply of 20 year-old secondary forests, many farmers are forced to reduce the fallow period, leading to soil degradation and lower crop yields. Farmers also depend on their woody fallows, secondary and primary forests for numerous tree products for local use or sale: construction material, firewood, food, medicine and fibres (Sotelo-Montes & Weber, 1997). For these farmers, therefore, an improved fallow must satisfy all the functions listed above.

Multistrata agroforests can also have several functions: maintain or improve soil fertility and physical properties, and control soil erosion; reduce weed invasion; sequester carbon in woody biomass, and reduce greenhouse gas emissions; diversify the production of crops and tree products on the same parcel of land; extend the production over a longer time, and increase the value of the land (Nair, 1984; Szott *et al.*, 1991) The sequential multistrata agroforest is the most common system in the Peruvian Amazon Basin: cropping cycle, secondary-forest fallow managed as multistrata, slash-and-burn, cropping cycle, etc. Very few farmers manage crops and trees in simultaneous multistrata agroforests or sequential Taungya systems because they perceive too many negative tree-crop interactions. On sloping land, a simultaneous multistrata system with contour hedges of various species between crops can control soil erosion, improve soil fertility and offer a diverse source of income (Alegre, unpublished data), but this system has not been widely adopted by farmers.

The International Centre for Research in Agroforestry (ICRAF) and collaborating research and development institutions in the Peruvian Amazon Basin are building upon these traditional land-use systems, and developing more productive, sustainable, biodiverse and dynamic agroforests at the landscape level. Two key components in these agroforests are improved woody fallows and enriched secondary-forest multistrata systems that satisfy all the functions listed above and provide additional social and environmental benefits. We hypothesize that widescale adoption of these key components will strengthen the rural economy, increase food security of resource-poor farming families and their domestic animals, stabilize the traditional slash-and-burn system, restore productivity in degraded lands, buffer the primary forest margin, decrease rates of deforestation, and help rebuild and conserve tree genetic resources at the farm, community and regional levels.

Several factors must be considered in designing improved woody fallows and multistrata agroforests. Which functions are most important for farmers and society at large? Which crop and tree species are most appropriate to satisfy these functions, considering not only the species' characteristics but also farmers' attitudes towards these species (e.g. desire to sow and manage long-rotation timber trees)? How do the component species and the whole system exploit environmental resources (light, water, nutrients)? What is the most effective spatial/temporal arrangement and management for the component species, considering the principal functions of the system, the species' biological characteristics, interspecific interactions, nutrient recycling and hydrology processes within the system, and the farmers' management practices, labour resources and market access for tree products? Considerable basic and applied research is still required to provide answers to these questions. To meet this need, ICRAF and collaborating institutions are investigating farmers' knowledge about agroforestry and agroforestry trees, and using the information to design, with farmer participation, "best bet" improved woody fallows and multistrata agroforests for process-oriented research and validation on farm.

FARMERS' USES OF *INGA EDULIS* IN IMPROVED WOODY FALLOWS AND MULTISTRATA AGROFORESTS

Following the cropping cycle, many farmers in the Peruvian Amazon Basin plant *Inga edulis* into their woody fallows (Brodie *et al.*, 1997; Potters, 1997; Sotelo-Montes & Weber, 1997). They harvest the fruit beginning in the second year for local consumption and sale: the edible part is the fleshy sarcotesta surrounding the seed. Although this is not a high-value fruit, it is the most widely traded native fruit in the region (Labarta-Chávarri & Weber, unpublished data), and is prepared commercially into juice and ice-cream flavouring. At the end of the fallow period, farmers typically harvest most of the woody biomass of *I. edulis* for firewood, and other selected species for construction material, firewood, charcoal, etc. Because nutrients in the harvested fruits and woody biomass are removed from the field, the farmers do not maximize the fertility-replenishment function of their woody fallows. Nevertheless, many of them recognize that the litter of *I. edulis* serves as a green manure during the fallow period. The use of *I. edulis* green manure for fallow improvement is well developed in some indigenous farming communities: they can maintain sustainable crop production for 5 years following a 5-year fallow enriched with selected *I. edulis* (Brodie *et al.*, 1997).

Many farmers in the region have access to expanding markets for timber and non-timber tree products, and they are enriching their secondary-forest fallows for these markets (Labarta-Chávarri & Weber, unpublished data). In the Pucallpa region, for example, they are selecting and sequentially planting *I. edulis* and other fruit trees (mainly *Bactris gasipaes*, *Citrus* spp., *Matisia cordata*, *Poraqueiba sericea* and *Spondias dulcis*) and timber trees (mainly *Calycophyllum spruceanum* and *Guazuma crinita*) in their secondary forests, with plans to manage them for long-term production (Brodie *et al.*, 1997). Fruit production in *I. edulis* declines after 6–10 years, so the farmers then harvest the above-ground biomass for firewood. They typically re-establish *I. edulis* by managing shoots from the coppiced stem, or resowing seeds. Farmers practice a very conscious selection of *I. edulis* based primarily on fruit-quality characteristics demanded in the Pucallpa market.

Nearly all farmers manage small home gardens, with *I. edulis* and other fruit trees, and a variety of plants for medicinal use and condiments in cooking. Indigenous communities seem to have developed more diverse home gardens than immigrant communities, and *I. edulis* is widely cultivated in these systems for fruit production, shade, firewood and green manure (Brodie *et al.*, 1997).

RESEARCH ON *INGA EDULIS* FOR IMPROVED WOODY FALLOWS AND MULTISTRATA AGROFORESTS

Some of the research on *Inga edulis* in the Peruvian Amazon Basin is summarized to illustrate the potential benefits of the species in improved woody fallows and multistrata agroforests. The research was conducted, or is underway, near Yurimaguas (2,200 mm annual rainfall, 180 m above sea level).

TABLE 1. Total biomass (above ground, litter, roots to 45 cm depth) of vegetation in a fallow enriched with *Inga edulis* and a natural secondary-forest fallow at various times after fallow initiation.

	Months in fallow					
Fallow	4	8	17	29	41	53
	Total biomass, t/ha dry weight					
Enriched with *I. edulis*	5.4*	20.0*	26.0*	43.0	57.0	72.5
Natural secondary forest	4.6*	6.0*	23.5*	37.0	52.5	69.0

Source: Szott *et al.*, 1994.
* Statistically significant difference between the fallows in the given month (P < 0.05).

A fallow enriched with *Inga edulis* can accumulate more biomass in the first few years than a natural secondary-forest fallow. Szott *et al.* (1994) evaluated biomass production in a natural secondary-forest fallow and a fallow enriched with *Inga edulis* on Ultisols. The secondary-forest fallows developed entirely from natural regeneration after the cropping cycle (referred to as natural fallow). *I. edulis* was sown at 2,500 plants per hectare into another field after the cropping cycle, where natural regeneration was also allowed to develop (referred to as enriched fallow). Total biomass (above ground, litter, roots to 45 cm depth) was measured at 4, 8, 17, 29, 41 and 53 months after initiating the fallows. The enriched fallow produced 3-times as much total biomass as the natural fallow during the first 8 months, continued to produce significantly more biomass at 17 months, but there were no significant differences after that (Table 1). The average annual increment in total biomass for the enriched fallow was approximately 16.4 t/ha, which compares favourably with other published values (Szott *et al.*, 1995). In both fallows, about 75–85% of the total biomass was above ground. The ratio of below- to above-ground biomass was similar to other published values for secondary-forest vegetation (ratio = 0.1–0.3): it was relatively constant over time, and did not differ significantly between the two fallows. Litter production, expressed as a percent of the above-ground biomass, increased during the first 17 months (~25% and 55% in the improved and natural fallow, respectively), remained at these levels through 41 months, and then decreased at 53 months (~15% and 20% in the improved and natural fallow, respectively), but there were no significant differences between the fallows (Szott *et al.*, 1994). Leaf area index (LAI) did not differ significantly between the fallows through 8 months, but it was considerably higher in the enriched fallow at 17 months (LAI 5.6 vs 2.3, P < 0.05). LAI was not reported beyond 17 months. Weed invasion was monitored for the first 29 months: weed biomass in the enriched fallow increased to 5.4 t/ha by 8 months, and then declined to 1.0 t/ha by 29 months (Szott, 1987), but there was no significant difference in weed biomass between the two fallow types.

Fallows enriched with *Inga edulis* can also increase the total stock of nitrogen, phosphorus and potassium in the system. In the experiment cited above, Szott & Palm (1996) recorded a 10% increase in the total stock (soil and vegetation) of N in the fallow enriched with *I. edulis* at 53 months, but no change in the natural fallow. Total stocks of P and K increased considerably during the same period (up 40% and 82%, respectively), while Ca and Mg stocks decreased (25% and 40%, respectively). The stocks of P and K decreased in the soil during this period, but this decrease was offset by uptake and accumulation of these nutrients in the living biomass and litter. Stocks of soil organic carbon decreased during the first 8 months due to high rates of mineralization, then increased slowly due to litter decomposition, but did not reach their initial levels by 53 months. Soil organic carbon was higher in the natural fallow than in the fallow enriched with *I. edulis* at 53 months (97% and 87.5% of initial values, respectively) because the litter of *I. edulis* decomposes slowly compared with most of the other species in the two fallows. Fallows enriched with *I. edulis* and leguminous cover crops may recover the initial level of soil organic carbon more rapidly than fallows enriched only with *I. edulis* (J. Alegre, unpublished data). From the farmers' point of view, a 24–36 month fallow period is clearly more desirable than a 53-month fallow period. Comparing the total stock of N, P, K, Ca and Mg at 29 months, the enriched fallow is superior to the natural fallow only for Ca (Table 2), apparently due to more rapid uptake and accumulation in the above-ground biomass and litter of *I. edulis*.

TABLE 2. Quantity of nutrients stored in above-ground biomass, roots (0–45 cm), litter, and soil (0–45 cm) in a fallow enriched with *Inga edulis* and a natural secondary-forest fallow after 29 months.

Fallow	Above-ground biomass	Roots	Litter	Soil	Total
	Quantity of nutrient, kg/ha				
Nitrogen					
Enriched with *I. edulis*	302*	93	106	4,860*	5,361
Natural secondary forest	153*	59	80	5,482*	5,774
Phosphorus					
Enriched with *I. edulis*	23	5	6	14	48
Natural secondary forest	15	4	5	16	42
Potassium					
Enriched with *I. edulis*	127	27	20	108	282
Natural secondary forest	176	22	16	122	336
Calcium					
Enriched with *I. edulis*	150*	27	62*	190	429*
Natural secondary forest	89*	16	39*	183	327*
Magnesium					
Enriched with *I. edulis*	27	12	14*	64	117
Natural secondary forest	30	10	11*	53	104

Source: Szott & Palm, 1996.
* Statistically significant difference between the fallows in the given month (P < 0.05).

Inga edulis, together with an herbaceous legume cover crop, will increase soil nitrogen levels more rapidly than a natural secondary-forest fallow or a pure *I. edulis* fallow (J. Alegre & L. Arévalo, unpublished data). This experiment, being evaluated on typic Paleudult soils, includes *I. edulis* with and without *Centrosema macrocarpum* (a legume cover crop), a natural secondary-forest fallow, and pure *C. macrocarpum*. *I. edulis* grew rapidly with minimal weeding during the first few months: after 2 years, the average tree height was 4.9 m and stem diameter at breast height was 5.5 cm. The cover crop quickly dominated the site without negatively affecting growth of *I. edulis*, produced abundant litter (3.2 and 5.1 t/ha in the first and second years, respectively), and effectively prevented weed establishment. In the second year, nitrate levels at 10 cm soil depth were twice as high under *I. edulis* with *C. macrocarpum* than under the natural secondary-forest fallow, apparently due to more rapid mineralization of their combined litter.

Contour hedgerows of *Inga edulis* on hillslopes can effectively reduce soil and water erosion, and maintain soil fertility on sloping lands (Alegre & Rao, 1995). This 6-year experiment, conducted on 15–20% slopes with typic Paeudult soils, included two treatments: cropping of rice (*Oryza sativa*) and cowpea (*Vigna unguiculata*) in annual rotation between *I. edulis* hedgrows arranged in contour lines (4 m between lines), and sole cropping of the rice and cowpea in annual rotation without the hedgerows. These were monitored for 15 consecutive crops. In addition, comparison plots were established in an adjacent secondary forest. The *I. edulis* hedgerow significantly reduced soil bulk density and increased saturated hydraulic conductivity, compared with the sole cropping treatment: bulk density = 1.29 vs 1.43 mg/m³ at 0–7.5 cm depth, saturated hydraulic conductivity = 50.0 vs 18.5 cm/hour (P < 0.05). When bulk density is high and saturated hydraulic conductivity is low, water infiltration is reduced and the soil is more susceptible to erosion, so the potential for erosion is reduced with the *I. edulis* hedgerow. Not surprisingly, the secondary forest had the lowest bulk density (1.20 mg/m³) and highest saturated hydraulic conductivity (99.8 cm/hour). The hedgerows conserved more soil and water in the system, compared with the sole cropping treatment: mean annual water runoff was 8-times greater, and soil loss was 59-times greater without the hedgerow. The hedgerow also maintained higher soil fertility compared with sole cropping: after 6 years, the hedgerow had 12% more organic carbon, 52% more phosphorus, 86% more potassium, 20% more calcium, 22% more magnesium (all P < 0.05), and 5% lower aluminium saturation (P < 0.01) than the sole cropping treatment. In addition, the secondary forest had more organic carbon than the hedgerow (13% more). Although the hedgerows conserved more water, soil and nutrients in the system, compared with the sole cropping treatment, this did not translate into striking increases in crop yields during the 6-year monitoring period: only 3 of 15 crop yields in the last 2 years were significantly higher in hedgerow intercropping than in sole cropping. Although a longer period of time may be necessary to realize the benefits of soil and water conservation, contour hedgerows of *I. edulis* are recommended for moderately sloping lands in the Peruvian Amazon Basin.

Inga edulis and other agroforestry tree species are being evaluated in a long-term study of a prototype multistrata agroforest (J. Alegre & L. Arévalo, unpublished data). The agroforest was established in 1985 as a Taungya system on an Ultisol, after slash-and-burn of a secondary-forest fallow. Rice and cowpea were interplanted with four tree species: *I. edulis* for fruit, firewood, and soil improvement; *Bactris gasipaes* for fruit and heart-of-palm, *Cedrelinga catenaeformis* for timber and soil improvement, and *Eugenia stipitata* for fruit. Both *I. edulis* and *C. catenaeformis* fix nitrogen and produce abundant leaf litter. Tree spacing was 5 × 5 m for *I. edulis* and *E. stipitata* (400 plants/ha), and 10×10 m for *B. gasipaes* and *C. catenaeformis* (100 plants/ha). After 2 years, the rice and cowpea were replaced with *Centrosema macrocarpum* for soil improvement and conservation, and to produce seed for improved pastures. The multistrata agroforests were established on two soils that differ in clay content (7% vs 17% in the upper 15 cm). They are being compared with various land-use systems on the same soil types, including a natural secondary forest that was 10 years old in 1985.

Inga edulis grew very rapidly in the multistrata agroforest and had to be pruned after 18 months to reduce competition with other plants. At that time, the canopy cover was 88% on the soil with more clay and 45% on the sandier soil (J. Alegre & L. Arévalo, unpublished data). The small branches were recycled as green manure in the agroforest, and larger branches were removed for firewood. The trees were pruned as necessary each year, and 50% of the trees were coppiced after 6 years.

TABLE 3. Biomass of different products removed from a multistrata agroforest during the first 10 years. Species included: Rice = *Oryza sativa*, Cowpea = *Vigna unguiculata*, Inga = *I. edulis*, Arazá = *Eugenia stipitata*, Peach palm = *Bactris gasipaes*.

Year	Rice t/ha	Cowpea t/ha	Inga fruit, t/ha	Inga firewood, m³/ha	Arazá fruit t/ha	Peach palm fruit, t/ha
1	2.82	0.90	—	—	—	—
2	0.77	—	1.10	—	—	—
3	—	—	3.29	62.2	1.22	—
4	—	—	—	—	2.10	—
5	—	—	0.67	—	3.48	—
6	—	—	—	23.9	0.73	10.30
7	—	—	1.05	24.0	—	6.75
8	—	—	0.25	—	—	6.20
9	—	—	—	—	—	5.45
10	—	—	—	—	—	9.30
Total	3.59	0.90	6.34	110.1	7.53	28.70

Source: J. Alegre & L. Arévalo, unpublished data.

Considerable biomass was removed from the multistrata agroforest during the first 10 years (J. Alegre & L. Arévalo, unpublished data). Mature fruit pods of *I. edulis* were stolen, so the number and weight of immature pods was determined. The estimated weight of mature pods and other products are given in Table 3. Although nutrients in the biomass were removed from the agroforest, the levels of soil nutrients (N, P, K, Ca, Mg) and organic carbon were similar to those found in the 20 year-old natural secondary forest (Figs. 1 and 2), presumably due to the effective nutrient recycling and nitrogen-fixation abilities of *I. edulis*, *C. catenaeformis* and *C. macrocarpum* in the

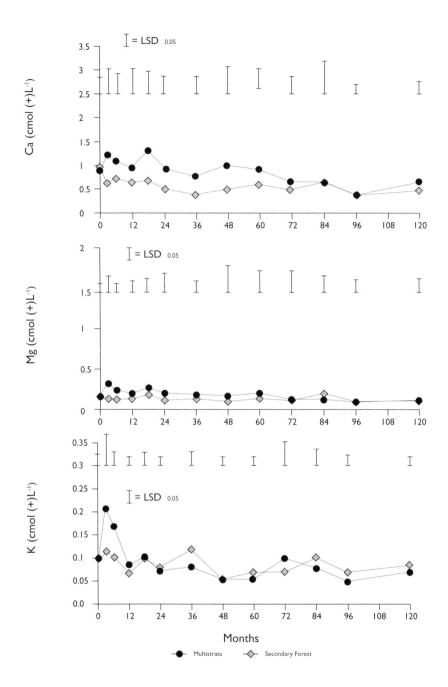

FIG. 1. Long-term dynamics of calcium, magnesium and potassium in the top 15 cm of the soil layer in the multistrata agroforest and natural secondary forest. Source: J. Alegre & L. Arévalo, unpublished data.

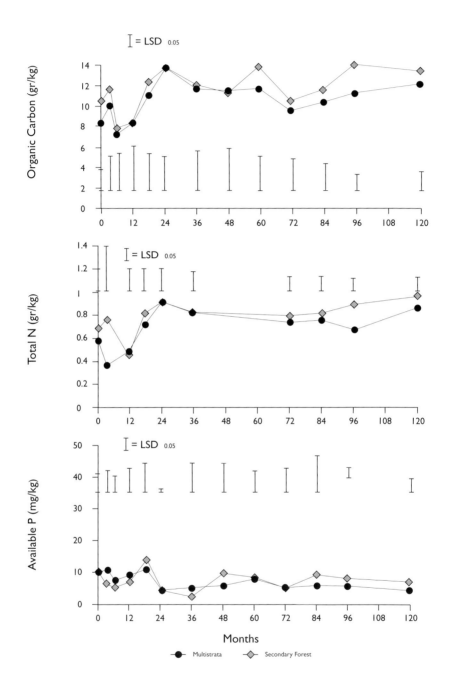

FIG. 2. Long-term dynamics of organic carbon, total nitrogen and phosphorus in the top 15 cm of the soil layer in the multistrata agroforest and natural secondary forest. Source: J. Alegre & L. Arévalo, unpublished data.

agroforest. In addition, soil acidity and aluminium saturation did not change much during the 10-year period in the agroforest or natural secondary forest, possibly because organic compounds in the litter were able to form complexes with the aluminium. At the beginning of the experiment, soils in the agroforest were slightly less acidic than those in the 10 year-old secondary forest (acidity = 1.15 vs 1.87 cmol (+)/litre, aluminium saturation = 45% vs 55%, respectively), and this difference increased slightly after 10 years (acidity = 0.91 and 2.00 cmol (+)/litre, aluminium saturation = 49% and 72%, $P < 0.05$).

The soils became more compacted in the multistrata agroforest over the 10 years (J. Alegre & L. Arévalo, unpublished data), probably related to the labour-intensive maintenance and harvest operations. Soil bulk density increased from 1.28 to 1.41 g/cm^3, and saturated hydraulic conductivity decreased by 50%. The increase in bulk density is actually a benefit on this particular soil: it increases the proportion of micropores relative to macropores, and thereby increases the water-holding capacity of the soil. On sloping lands, *Inga edulis* could be established in hedgerow contour lines to minimize soil and water erosion in the agroforest.

SELECTION OF *INGA* SPECIES FOR IMPROVED WOODY FALLOWS AND MULTISTRATA AGROFORESTS

The genus *Inga* contains approximately 300 species in the neotropics, and the majority of them occur in the extensive Amazon River Basin (Pennington, 1997). There is very little published information about the ecology, management, production potential, intraspecific variation, etc. for most of these species. How can we select the best species of *Inga* for agroforestry research and development? ICRAF and the International Service for National Agriculture Research (ISNAR) developed a methodology to prioritize agroforestry tree species based on farmers' preferences, knowledge and market access, and other factors (Franzel *et al.*, 1996). Farming communities which cultivate *Inga* species have considerable knowledge about these species (Lawrence, 1995; Brodie *et al.*, 1997; Potters, 1997). Documentation, systematization and dissemination of this local knowledge could help research and development organizations to better identify priority species, and improve the management and use their genetic resources. ICRAF is leading a comprehensive study of farmers' knowledge about agroforestry tree species, in collaboration with the International Centre for Tropical Agriculture (CIAT), the National Institute for Agricultural Research (INIA) of Peru, and the Confederation of Amazon Nations of Peru (CONAP). The information will be published in a new database for distribution.

Peru is one of the richest centres of *Inga* diversity, with 91 recognized species (Reynel & Pennington, 1997). From 1989 to 1991, researchers from various national and international institutions started a project to evaluate *Inga* species for improved woody fallows, multistrata agroforests and other agroforestry systems. They collected seeds from 303 mother trees in 14 locations in the Peruvian Amazon Basin, and established field trials in Yurimaguas, Peru to evaluate inter- and intraspecific variation. T. D. Pennington (The Royal Botanic Gardens, Kew) and O. Poncy (National Museum of France) identified more than 75% of the accessions, using botanical samples collected in the field trials. These include 30 species, some previously unrecognized. Although several species in the field trials are represented by only a few progeny families, there is clearly a lot of interspecific variation in growth rate, fruit and firewood production, response to pruning and coppicing, and other characteristics relevant to improved fallows and multistrata agroforests (L. Szott, R. Ricse & J.C. Weber, unpublished data). In the case of *I. edulis*, which is well represented in the field trials, there is also significant variation in growth rate and biomass production among progeny families (Szott *et al.*, 1995). Lawrence *et al.* (1995) also reported considerable variation among *Inga* species collected from Ecuador, Costa Rica and Honduras, but did not investigate intraspecific variation.

ICRAF and the National Agricultural Research Institute (INIA) of Peru are working with farmers to select and multiply superior germplasm of *Inga edulis* and other species on farms in the Peruvian Amazon Basin (J.C. Weber, principal investigator). In the case of *I. edulis*, farmers identify their best trees for (1) fruit, (2) biomass and (3) green manure, and explain their criteria for selection to the researchers. Seeds are collected from the trees selected by farmers, and are used to establish replicated progeny tests for (1) fruit, (2) biomass and (3) green manure within several watersheds. At least 20 farmers participate in each watershed, and each farmer manages one replication in collaboration with researchers. The participating farmers are organized into informal networks to facilitate exchange of information and germplasm within each watershed. Farmers and researchers jointly evaluate the germplasm so that (1) researchers understand farmers' selection criteria and threshold for adoption of improved germplasm, (2) researchers document farmers' knowledge about the species, and (3) farmers' benefit from technical advice of researchers. Participating farmers receive all products harvested from the progeny tests. Based on the evaluation results, superior progeny families and trees within families are selected in each replication after 3 years, and less promising germplasm is coppiced. The replications are then managed as seed orchards for production of improved germplasm at the farm and watershed levels, and participating farmers receive income from the sale of the improved germplasm.

ICRAF and collaborators believe that conservation-through-use of selected germplasm of *Inga edulis* and other agroforestry tree species will have several benefits. It will increase genetic diversity and value within the species in each watershed, following the multiplication, distribution and exchange of the improved germplasm. Exchange of improved germplasm among networks from different watersheds will augment the genetic diversity and value at a regional level. Use of the improved germplasm will increase productivity and sustainability of agroforestry systems such as woody fallows and multistrata agroforests, reduce rural poverty and food insecurity, and reduce deforestation pressures on remaining primary forest.

References

Alegre, J.C., Cassel, D.K. & Bandy, D.E. 1986. Effects of land clearing and subsequent management on soil physical properties of an Ultisol in the Amazon Basin of Peru. Soil Sci. Soc. Amer. J. 50: 1379–1383.

Alegre, J.C. & Rao, M.R. 1995. Soil and water conservation by contour hedging in the humid tropics of Peru. Agric. Eco-Syst. Environm. 57: 17–25.

Brodie, A.W., Labarta-Chávarri, R.A. & Weber, J.C. 1997. Tree germplasm management and use on-farm in the Peruvian Amazon: a case study from the Ucayali region, Peru. Research report, Overseas Development Institute, London and International Centre for Research in Agroforestry, Nairobi.

Denevan, W.M. & Padoch, C. (eds.). 1990. Agroforestería Tradicional en la Amazonia Peruana. New York Botanical Garden and Centro de Investigación y Promoción Amazónica, Lima.

Franzel, S., Jaenicke, H. & Janssen W. 1996. Choosing the Right Trees: Setting Priorities for Multipurpose Tree Improvement. ISNAR Research Report 8. The Hague: International Service for National Agricultural Research.

ICRAF. 1995. Alternatives to slash-and-burn. In: Annual Report 1995, International Centre for Research in Agroforestry, pp. 187–196. ICRAF, Nairobi.

Lawrence, A. 1995. Farmer knowledge and use of *Inga* species. In: D.O. Evans & L.T. Szott (eds.), Nitrogen Fixing Trees for Acid Soils, Nitrogen Fixing Tree Research Reports (Special Issue), pp. 142–151. Winrock International and Nitrogen Fixing Tree Association, Morrilton, Arkansas.

Lawrence, A., Pennington, T.D., Hands, M.R. & Zúniga, R.A. 1995. *Inga*: High diversity in the neotropics. In: D.O. Evans & L.T. Szott (eds.), Nitrogen Fixing Trees for Acid Soils, Nitrogen Fixing Tree Research Reports (Special Issue), pp. 130–141. Winrock International and Nitrogen Fixing Tree Association, Morrilton, Arkansas.

Leakey, R. 1996. Definition of agroforestry revisited. Agroforest. Today 8(1): 5–7.

Nair, P.K. 1984. Soil Productivity Aspects of Agroforestry. International Council for Research in Agroforestry. Nairobi, Kenya.

Nye, P.H. & Greenland, D.J. 1960. The Soil Under Shifting Cultivation. Commonw. Bur. Soil, Techn. Commun. 51, Harpendon, England.

Pennington, T.D. 1997. The Genus *Inga* – Botany. The Royal Botanic Gardens, Kew.

Potters, J. 1997. Farmers' Knowledge and Perceptions About Tree Use and Management – the Case of Trancayacu, a Peruvian Community in the Amazon. M.Sc. Thesis, Department of Forestry, Wageningen Agricultural University, Wageningen, The Netherlands.

Reynel, C. & Pennington, T.D. 1997. El género *Inga* en el Perú, Morfología, Distribución y Usos. The Royal Botanic Gardens, Kew.

Sanchez, P.A. 1976. Properties and Management of Soils in the Tropics. Wiley, New York.

Sotelo-Montes, C. & Weber, J.C. 1997. Priorización de especies arbóreas para sistemas agroforestales en la selva baja del Perú. Agroforestería en las Américas 4 (14): 12–17.

Szott, L.T. 1987. Improving the productivity of shifting cultivation the Amazon Basin of Peru through the use of leguminous vegetation. Ph.D. Dissertation, North Carolina State University, Raleigh, N.C.

Szott, L.T., Fernandez, E.C.M & Sanchez, P.A. 1991. Soil-plant interactions in agroforestry system. Forest Ecol. Managem. 45: 127–152.

Szott, L.T., Palm, C.A. & Davey, C.B. 1994. Biomass and litter accumulation under managed and natural tropical fallows. Forest Ecol. Managem.67: 177–190.

Szott, L.T., Ricse, A. & Alegre, J. 1995. Growth and biomass production by introductions of *Inga edulis* in the Peruvian Amazon. In: D.O. Evans & L.T. Szott (eds.), Nitrogen Fixing Trees for Acid Soils, Nitrogen Fixing Tree Research Reports (Special Issue), pp. 237–249. Winrock International and Nitrogen Fixing Tree Association, Morrilton, Arkansas.

Szott, L.T. & Palm, C.A. 1996. Nutrient stocks in managed and natural humid tropical fallows. Pl. Soil 186: 293–309.

CHAPTER 7. *INGA* AS SHADE FOR COFFEE, CACAO AND TEA: HISTORICAL ASPECTS AND PRESENT DAY UTILIZATION

JORGE LEÓN

The use of shade trees in the culture of perennial crops in the tropics, especially in cacao and coffee and to a lesser extent in coca and tea, arose independently in the Old and the New World, at different places and times, evidently as a response to some local problems. Once the practice was established in a region, it spread quickly to other areas, although it was never a practice of universal acceptance.

Shade has multiple and complex effects on the crop environment, affecting the quantity and quality of light, soil conditions, air and soil temperature, and incidence of diseases and pests. It affects the income of the farmer, as it decreases yield. However, as coffee and cacao prices change according to factors completely outside the influence of the farmer, the control of the shade allows him to adapt production to market conditions.

An argument in favour of shading coffee is that in its native habitat, the highland forests of Ethiopia, *Coffea arabica* grows under heavy shade provided by tall trees. The observations of scientists and travellers is that under such conditions the yield is very low, as can also be observed at any site in which coffee is grown under heavy shade. As in other crops, the domestication process implies changes, often drastic, of the environmental conditions of the native habitat. Shade may be provided by other trees in the traditional system of coffee growing, or by the coffee bushes themselves in the modern system. The growth of abundant foliage is supported in the latter system by heavy fertilization, irrigation and the control of diseases and pests.

The practice of shading perennial crops has always had detractors and followers. Most of the arguments in favour or against the practice are based on personal experiences or preferences, tradition or rational interpretations. The experimental approach is complex to apply, due to the difficulty of replicating artificially the effects and interactions of the many factors involved in shade effect. This is evident in the attempt to measure solar radiation intensity under artificial conditions (Guiscafré-Arrillaga, 1942).

The first attempt to analyse the pros and cons of shade in coffee production was made by Cook (1901), who was opposed to it. By the same time the eminent agronomist H. Semler (1892–1900) stated that shade had been adopted as a general practice, but he suggested that it is unnecessary above 1100 m, but needed in lower areas, especially in dry climates and at wide spacing. This is, more or less, the present opinion of coffee specialists.

An interesting discussion promoted by Cook's publication, was held in Costa Rica (Mora, 1910; Pérez Zeledón, 1910; Van der Laat, 1910) and although these three authors were against shade, a farmer replied in favour of its use in a series of newspaper articles. A quarter of a century before, in Colombia, several farmers had discussed the subject on a practical basis, and came up supporting the use of shade (Patiño, 1969).

The agronomic aspects of shade have been revised several times (Ostendorf, 1962; Willey, 1975; Kimenia & Njoroge, 1988; Fournier, 1988).

The situation in Brazil, where almost all coffee is planted without shade, differs from the rest of the tropical American countries. This can be explained by the difference in the water available in the soil during the dry season. According to Franco (1957), the first coffee planted in São Paulo was under shade, but this practice had to be abandoned because of low yields. The experimental trials show that in plots shaded with *Cassia strobilacea* and without shade, the amount of water in the soil was less in the first. In another trial, using *Inga* as shade, and measuring evapotranspiration, the deficit for the unshaded plot was 25 mm while in the shaded plot it was 274 mm.

In the discussions on shade, the social and economic aspects of the problem have received practically no attention. The publications refer to yields, without analysing costs and net incomes, at farmer and country levels. A remarkable exception is the work of Ridler (1982).

The choice of shade trees

First it is necessary to distinguish the trees planted to provide shade from those interplanted with the intention of obtaining an additional product: fruit, timber, rubber.

Since the beginning of planting shade trees, the first choice were leguminous species. Nevertheless, it is not any kind of leguminous species that can be used, and Cramer (1957) reported that coffee growth and yields declined when planted close to *Cassia florida*. The large majority of shade trees belong to the *Leguminosae*: *Mimosoideae*: *Albizia*, *Inga*, *Leucaena*, or to the *Leguminosae*: *Papilionoideae*: *Erythrina*, *Gliricidia*. As in the discovery of the value in soil improvement, reflected in better growth or yield, of crops associated with pasture legumes in the ancient agriculture in the Old World, empirical observations of improved growth and yields of shaded crops by the Mesoamerican Indians led to the use of *Gliricidia sepium* as cacao shade.

This discovery was based, presumably, on the observation that cacao plants growing close to *Gliricidia sepium* had better growth, more foliage and lived longer than in other situations. Eventually the same results were observed when cacao grew close to *Erythrina* and *Inga* plants. The farmers were of course, not aware of the nitrogen fixing ability of these species, but they observed the results of this process (Budowski *et al.*, 1984).

There is no basis to support the possibility that the use of shade in coffee and other crops in the New World, originates in the practice of shading cacao with *Gliricidia sepium* in Mesoamerica.

The use of leguminous trees in shading perennial crops was independently discovered also in SE Asia, as may be inferred from the recommendation of the Dutch East Indies Company, in 1795, to plant *Erythrina lithosperma* between coffee trees (Cramer, 1957).

The effects of nitrogen uptake by crop trees interplanted with leguminous trees was based on indirect evidence, even when nodulation was observed in the shade trees. In fact, it was only recently that this effect was measured, in coffee interplanted with *Inga jinicuil* (Roskoski, 1981; Fernandes, this volume). Several aspects of the nodulation process need to be studied, one is the possible decrease with age, as is suspected by farmers; the nodulation differences according to the species; the relationship between nodulation and mineral fertilization as found by Roskoski (1981) and other problems.

Another source of nitrogen and its release by the mineralization of the leaves in the soil, is being studied in several species of *Inga* (Palm & Sanchez, 1990; Meléndez *et al.*, 1995).

Many species of trees, leguminous or not, have been used to shade tree crops in the tropics (Cook, 1901). The trend, however, has been to reduce their number and to concentrate on a few species in two of the three subfamilies of the *Leguminosae* (*Mimosoideae* and *Papilionoideae)*. Of the third subfamily, *Caesalpinoideae*, only *Cassia grandis*, is being planted and reported to be a good shade tree, resistant to wind (Montealegre, 1938). There is no information on nodulation in this species, and it is well known that nodulation is scarce in the *Caesalpinoideae.*

Shade trees are a sort of semicultivated plant. They are chosen on an array of different characters, such as shape of the canopy, size, duration, resistance to diseases and pests, soil relationships and others (Alvarado, 1935; Beer, 1987). They are subjected to horticultural practices: planting, pruning, vegetative propagation, seed reproduction of selected progenies. As in other cultivated plants, they are changed or replaced by other species or varieties, often introduced from other countries. The knowledge of their management is empirical, and it is transmitted orally among farmers. In spite of its practical importance, no systematic research has been made on their selection and husbandry, except in *Erythrina poeppigiana.*

In *Inga*, due to the large number of species and their difficult identification, the information is not as reliable as in *Erythrina*, a genus in which only half a dozen species are planted as shade trees. In *Inga* several vernacular generic names are used, often followed by specific epithets. Thus, ingá is the Brazilian name for the genus; inga-cipó is *I. edulis*; ingapeu is *I. macrophylla*; inga-turi, *I. alba*, and so on. Other generic names are guamo and its derivatives: guabo and guamá; *I. edulis* is guabo mecate; *I. fagifolia* guamá. The name guamo comes from an Indian dialect spoken in Hispaniola; it is used in the Antilles (Marrero, 1954), Colombia (Anon., 1958) and Venezuela (Escalante *et al.*, 1987). A nahuatl name, cuajiniquil, is used from Mexico to Costa Rica for several species of *Inga*, as pacae is used in the Andean region in South America. There are many local names, derived from Indian dialects or European languages, of restricted distribution.

The most common species of *Inga* used as shade trees can be differentiated on fruit characters.

1. The first group is characterised by cylindrical fruits, often twisted. These are probably the most appropriate trees for shade purposes (Uribe, 1945). *Inga edulis* is remarkable for the architecture of the plant: a low trunk with few branches forming and umbrella-like canopy; leaves of different size and shape according to variety (León, 1966); rapid growth; long life, with trees of 30–50 years still in good condition. Some varieties produce good fruits, a disadvantage for shade purposes, others are inedible. Common names: guabo mecate (Central America), guabo rabo de mono, guabo santafereno (Colombia), guamo mecate (Venezuela), ingá cipó (Brazil).

The closely related *Inga oerstediana* has a wide altitudinal range, 0–1800 m, and thus is used to shade cacao and coffee. It has a tall, wide and open canopy; the large leaves make an excellent mulch; the fruits are worthless. In the coffee literature the species mentioned as *I. portobellensis* is probably *I. oerstediana*. It is susceptible to insect and fungus attacks (Montealegre, 1938). Common names: cushin (Guatemala, El Salvador), cuajiniquil peludo (Costa Rica).

Inga vera (including *I. eriocarpa, I. spuria*), is found from Mexico to Brazil and in the Greater Antilles. It is a fast growing tree, with spreading branches, resistant to wind, excellent for renovating old coffee fields (Alvarado, 1935). However, it has serious insect pests. The fruits are inedible. Common names: acotope (Mexico), shalum (Guatemala), pepeto real (El Salvador), cuajiniquil (Mexico–Costa Rica), guamo arroyero (Colombia), guaba del país (Puerto Rico), guamá (Dominican Republic), guabo (Central America, Cuba), guamo bobo (Venezuela), pacae (Peru).

2. Among the *Inga* species with small, flat fruits, *Inga fagifolia* (= *Inga laurina*) is planted as a shade tree in Colombia and the Antilles. It is a low tree, with a compact canopy. The small pods are edible. It has serious pests (ants) and it is attacked by root diseases. Common names: guamo rosario (Colombia), guama (Puerto Rico), jina (Dominican Republic), pois doux (Antilles), Spanish oak (Antilles).

Inga marginata is seldom used as coffee and cacao shade; it grows from Costa Rica to Brazil; it is a compact tree, with shiny foliage, fruits inedible. The root system is shallow and thick, with plenty of nodules. Common names: cuajiniquil negro (Costa Rica), guamo chirimo (Colombia), guamo caraota (Venezuela).

Inga punctata (syn. *I. leptoloba*) is found from Mexico to Brazil and Trinidad and Tobago. It is sporadically used as shade for coffee and cacao, but in some regions of Mexico, Honduras and Venezuela it is frequently planted. Common names: chalahuite (Mexico), pepeto negro (El Salvador), guamo caraota (Venezuela).

3. Of the *Inga* with quadrangular fruits, there are three species. *I. feuillei* is known only in the cultivated state, from Colombia to Chile. It is a tall tree, with an umbrella-like canopy, leaves softly pubescent, fruits up to 40 cm long, the seeds covered with a white, sweet pulp. Common name: pacae (Colombia to Chile). *Inga sapindoides* grows from Mexico to Brazil, and in Trinidad and Tobago. It is a low tree, with irregular canopy, pilose leaves, fruits 10–30 cm long, with very little pulp. Its use as shade is decreasing. Common names: tamatama (Belize), guabo cuadrado (Costa Rica), guabo cajeta (Venezuela). *Inga striata* was formerly used in south east Brazil (Castro, 1952).

4. Among the *Inga* species with flat, large and broad fruits, it is possible to distinguish two kinds: the first has flowers in short spikes; in the second, the flowers form umbels or spheres. *Inga densiflora* belongs to the first group; it grows from Costa Rica to Brazil, and it is used for shade in Central America and Colombia. It is a rather low tree, with an irregular canopy, with leaves

slightly pilose. The flat fruits, up to 30 cm long, are sold in the markets of Colombia and Costa Rica. In Colombia this species has been called *I. langlassei*, *I. microdontha*, *I. sordida* and *I. tiribiana*; in Venezuela, *I. heinei* and *I. java*. Common names: guabo salado (Costa Rica).

The *Inga* species with spherical or umbrella like inflorescences include *I. jinicuil*, from Mexico to Ecuador (Pennington, 1997), planted for shade and fruits. It is a rather low tree with compact canopy and shiny foliage. The fruit, flat and thick, up to 25 cm long, contains a sweet white pulp surrounding the seeds. Common names: jinicuil (Mexico), paterno (Central America). Its use as a shade tree is waning, as the trees are affected by a witches broom disease, and by the damage to coffee done by people who collect the fruits. *Inga nobilis* ssp. *quaternata* grows from Mexico to Venezuela; it is a low tree, with a thick canopy, leaves yellow green, slightly shiny, fruits flat, up to 20 cm long. It has been introduced to Puerto Rico, where it is known as *I. speciosissima*. Common names: acotopillo (Mexico), guabo cansa boca (Panama), guama venezolano (Puerto Rico).

THE CASE OF ARABICA COFFEE

Arabica coffee is the most important species in coffee production and trade. As it is grown mainly in Latin America, the following discussion on the development of shade systems, the trend to eliminate them and the replacement by agroforestry practices, is limited to the New World. It is also the only area where *Inga* spp. are used on a large number of farms.

It is quite difficult to document the early practices of coffee growing in the New World. There is almost a complete lack of written references on the development of cultural practices and their spread from one country to another. The introduction of coffee to the New World, on the contrary, is rather well known (Chevalier, 1929). One plant grown in the botanical garden at Amsterdam from seed received from Java between 1706–1710 was sent to Paris in 1713. From this plant, seeds were sent to the French colonies in America: Guadeloupe, French Guyana, Haiti and Martinique, between 1720–1724.

Commercial coffee production was developed in Haiti, and it was in this country that the practices for growing and processing were developed (Wellman, 1961). According to Moral (1955), there are a few references on coffee husbandry in Haiti during the eighteenth century. In 1778, a book was published by a French planter, J. Laborie, with information for Jamaican farmers on how to grow coffee, following the practices current in Haiti. Neither in the paragraphs copied by Moral (1955) nor in the design of a plan for a coffee plantation, included in the book, is there any reference to shade. Moral (1955) quotes other books by travellers in Haiti during the eighteenth century, but in their references to coffee growing there is nothing about shade. Wellman (1961) mentions coffee trees "grown in the shaded garden of the Terre Rouge" in Haiti, but he does not quote the source of this information. Moral (1955), on the contrary, comparing the present systems of coffee "plantations" in Haiti with the systems prevalent in the eighteenth century, states that in the latter "le café se plantait...à decouvert".

The coffee industry in Haiti, by that time the most important in the world, was ruined by the revolutions and the consequences of independence. The French planters first and later the Spanish that lived in Santo Domingo, had to migrate to Cuba (Pérez de la Riva, 1944) or to Venezuela (Baralt & Diaz, 1939), where they established new plantations according to the practices developed in Haiti.

Cuba received the first coffee seeds from Haiti in 1748 (Pérez de la Riva, 1944), but commercial production started with the arrival of the French planters from Haiti (Moral, 1955). One of them, Alexandre B.C. Dumont, published a book in 1833 on the cultural practices for coffee in Cuba. It contains all sorts of minute instructions, but nothing on shade.

From Cuba, coffee seeds were sent to Costa Rica at the end of the eighteenth century. Costa Rica established the first commercial production in Central America. The Dumont book was reprinted in Costa Rica in 1835. Some years later, a Costa Rican planter wrote a book (Aguilar, 1845) for the Guatemalan coffee farmers, on the system followed in Costa Rica for growing and processing coffee. As in Dumont's book, there are all sorts of details on cultural practices but nothing on shade. A painting, by an anonymous artist, of the landscape nearby San José, Costa Rica, done around 1850, shows the rows of coffee bushes but no shade trees.

As mentioned before, the coffee industry of Venezuela was established by French planters, and from Venezuela it spread to Colombia (Patiño, 1969). The development of the coffee industry of Colombia, the second in the world, started in the second half of the nineteenth century. There is no information available on the use of shade in the early stages of the coffee industry. In 1880, M. Ospina, in a handbook for coffee growers, states his doubts on the advantages of shade, and he recommends trying both systems, with and without shade (Ospina, 1952). From this, it may be surmised that shadeless coffee could have been planted in Colombia. In Guatemala, Preuss (1901) found that above 900–1000 m most of the plantations were shadeless, and in the new ones, the only shade was provided by a few wild trees that were left standing when the forest was cut. This system is documented also by the series of photographs taken some years earlier (Bradford Burns, 1986).

THE BEGINNING OF THE USE OF SHADE

It is likely that the practice of shading coffee started in Cuba. As mentioned above, coffee production at a commercial level began after the French and Spanish farmers left Haiti and started plantations in Cuba at the beginning of the eighteenth century. In 1790 Cuba exported 1,850 bags of 100 lbs; in 1805, 17,400; in 1815, 230,000 (Ely, 1962).

However, the industry had serious problems in the second and third decades of the nineteenth century, including lack of slave labour, high interest rates, price fluctuations, hurricanes and others. Price fluctuations were of high impact; in 1808 the price of the 100 lb bag came down from US$30 to US$3 (Ely, 1962). The wars between France and Spain forced many French farmers, the best coffee growers, to leave Cuba. The coffee exports declined from a maximum of 641,597 bags of 100 lbs in 1833 to an average of 50,660 in 1857–59. By 1860, Cuba was importing coffee from Brazil (Pérez de la Riva, 1944). The situation of the coffee industry at that time was discussed in meetings, books and articles. Many coffee fields were abandoned; there was no fertilization or weed control, and the rising sugar cane industry attracted capital and required more labour. Balmaseda (1890) in a book published many years later, described the situation of the coffee farms. He refers to two systems: the French consists of clearing the land and planting the coffee bushes exposed to the sun, poorly protected only by bananas and oranges that grow in the borders of the fields. In the other less common system, the forests were cleared of shrubs and the high trees pruned of their larger branches to permit enough light to reach the coffee bushes. The author concludes that in hot regions, the French system had disappeared due to the lack of shade. In 1840, shade was being established, as a newspaper in Santiago de Cuba advertised the sale of bucare seeds, *Erythrina* sp. (Pérez de la Riva, 1944).

Socioeconomic factors explain in part the decrease of coffee production in Cuba, and there is data to support this. On the agronomic aspects, however, practically nothing is known. After two decades of high production based on the natural resources of the soil, with poor and erratic fertilisation, inadequate pruning and very likely with high erosion and loss of organic matter, a continuous decrease in productivity had to be expected. In Indonesia, after many years of discussing the low and erratic yields of arabica coffee, they were attributed to soil and climatic conditions (Cramer, 1957).

No information is available on how the practice of shading trees spread to other countries. In Costa Rica, according to Mora (1910), the use of shade was introduced from Puerto Rico in 1865. In Colombia, in 1872, farmers were advised to shade coffee with *Erythrina* in the lowlands and with *Inga* in the highlands (Patiño, 1969). By the end of the century, the practice had spread throughout the country (Sáenz, 1895). In Jamaica, the species recommended was pois doux, an *Inga* (Lock, 1888) and in Guadeloupe, *I. laurina* (Lecompte, 1899). In Colombia, Ospina in 1880 recommended *Erythrina*, *Albizia carbonaria*, and especially *Inga edulis* (Ospina, 1952). Around 1890, in Costa Rica (Pittier, 1929), *Inga* spp. were displacing *Erythrina* spp. as the favourite shade trees.

Once the practice of shading trees was accepted by farmers, the trend was to plant it in excess, based, very likely, on the principle that the more the better. The management of the shade trees was unknown, especially the pruning practices; some provisional shade, like bananas, were often left for economic reasons. The condition of the coffee groves deteriorated in the first decades of the twentieth century, due not only to the excess of shade but to other factors, such as lack of fertilization and low prices, but the shade was blamed above all of them. There were many discussions on the advantages and disadvantages of shade, based on rational arguments and empirical experience, as may be seen in Cook's book (1901). Perhaps the most serious drawback of the use of shade was to diminish the effect of chemical fertilization.

The depletion of soils in Brazil in the coffee areas without shade, strengthened the support of the shade system in other countries. Also the measurements done in Puerto Rico and Colombia in coffee plots shaded with *Inga*, showed that soil erosion and water run-off were minimal.

107

It was the Hawaiian experience in growing coffee without shade that moved some coffee growers in Latin America to start experimenting with this practice. They began by removing all the shade in plantations already established, with catastrophic results. It was the hedge system without shade, started in Guatemala (Cowgill, 1954), that eventually developed into the modern technology of coffee production.

SHADE IN THE NEW COFFEE TECHNOLOGY

The main characteristic of the modern system of coffee production is the absence of shade trees. This feature has completely changed the landscape of the coffee production areas in large areas, from Mexico to Colombia. Besides the elimination of the shade trees, the system includes hedge planting, that partially replaces the shade formerly supplied by trees; new coffee genotypes of compact growth; heavy applications of fertilisers and chemical control of weeds, pests and diseases. Although in many places the hedge rows are planted on contour lines, there is high soil erosion, and some means of control are being applied.

The new technology produces high yields and earlier crops, but demands an increase in labour, often difficult to obtain. The modern system is practised in areas of optimum growing conditions, at the appropriate range of temperature and rainfall, and on soils of high natural fertility and excellent physical conditions. Most of these areas were already occupied by large coffee enterprises, that can afford the high expenses demanded by the intensive use of fertilisers and other chemical products. However, numerous small farms located in the areas of optimum conditions, are now planted in the new system.

There is no reliable data on the proportion of land occupied by farms applying the modern system in the coffee producing countries, but in most of them it seems to be less than one fourth of the area planted with coffee, with the possible exceptions of Colombia and Costa Rica. Most of the farms that are still working with the traditional system are located in marginal areas to those having the optimum conditions, and the coffee growers in them do not have the economic and technical resources demanded by the new technology. For them, the advice given by the extension agencies, is to maintain the shade to save on fertilisers, but to plant shade trees at a lower density. There is also advice on changing the species used for shade. As the majority of the *Inga* species are susceptible to attack by insects, borers or defoliators (see Ackerman *et al.*, this volume), except a few species like *I. densiflora*, the recommendation is to change to *Erythrina*, particularly *E. fusca* and *E. poeppigiana*. The use of the latter species is increasing and it is the most common in experimental trials.

The evaluation of different species of *Inga* and their resistance to pests and disease, their response to pruning practices and their adaptation to drier habitats, urgently requires a good research and extension programme.

The successful results obtained by farmers pollarding trees of some *Inga* species in El Salvador (Lawrence & Zúniga, 1996) requires experimental testing, and may open the possibility of incorporating *Inga* shade trees in the modern systems of coffee production.

PROSPECTS

At the moment, the use of *Inga* as shade trees is in a chaotic situation. Many species are used, often only in reduced areas; very little is known on their husbandry, their capability of fixing nitrogen, and their inter-relationships with the new coffee and cacao genotypes.

Contrary to what is happening with *Erythrina*, very little research is being done on the practical aspects of the biology of *Inga*. The research in this field has to concentrate on a few species, perhaps no more than five, and on such aspects as the evaluation and management of branching patterns, as the factors determining planting distances and pruning practices. A second aspect is the resistance or tolerance to insects and foliar diseases, and the role of each in their spread to cacao and coffee plants. In the third place, the rates of nitrogen fixation and soil mineralization. It is possible that no single species of *Inga* fulfils these three conditions at a satisfactory level. This brings up the old idea of planting several species in the same field; to evaluate the pros and cons of these systems would require extensive and complex research.

One conclusion is clear in reviewing articles on shade species; it is that *Inga edulis* is the one species that seems to be the best choice. There is an urgent need to know more about this species, especially in the differences in size of leaflets and branching patterns of the different genotypes. The use of this species should be spread to areas in which it is unknown at the present time.

The future of *Inga* as shade depends in a large part on the utilization of *Erythrina* spp. with the same purpose, especially *E. poeppigiana*. This species is now widely used in restoring shade in the coffee plantations established without shade. It is also the species most frequently used in research trials, and some aspects of its management are well known, such as the vegetative propagation by large cuttings, which permits the establishment of uniform stands that supply shade the first year after planting. Another advantage is the pruning system, in pollards that make it easier to maintain an uniform stand.

Coffee, and to a lesser extent cacao, are planted without shade in areas subject to special environmental and social conditions, where large inputs of fertilisers, other chemicals and labour are available. But contrary to the dictum of some specialists that coffee should not be planted where is does not grow without shade (Fukunaga, 1957), coffee continues to be produced under shade, in conditions that are satisfactory to the farmer and that may be improved by research and extension. The traditional systems require less inputs at the farmer level and represent a large saving for the producing countries, as most of the inputs have to be imported. In Costa Rica "coffee at full sun scarcely produces 10% more than in conditions of balanced shade, with the disadvantage that coffee in full sun is more attacked by *Cercospora coffeicola*, and that the large quantity of weeds increases the cost of its control" (Pérez Solano, 1983).

In the new approaches in agroforestry, the leguminous shade trees have a special role (Budowski *et al.*, 1984, Phillips-Mora, 1983). Other general uses of *Inga* are discussed by Lawrence (1995) and Lawrence *et al.* (1995), with emphasis on their possibilities in alley cropping and the production of biomass.

The improvement of the traditional systems using *Inga* and other shade trees could mean not only increases in production at reasonable cost, but less losses of soil and less damage to the environment (Perfecto *et al.*, 1996). A comparison between organic and high technology production of coffee in Costa Rica, shows less total income in the first, but higher net revenue in the second (Ackerman & van Baer, 1992; Boyce *et al.*, 1994).

A final consideration is that coffee and cacao are marketed more and more on quality or on speciality types. In coffee it has been shown that the coffee produced under shade is of superior quality, no matter which shade tree is used. Some of the finest Central American coffee is grown under *Eucalyptus*. In certain markets, organic coffee, which requires shade, commands higher prices.

SHADE IN CACAO

Shading cacao was a regular practice in Mesoamerica before the arrival of the Europeans. At that time, cacao was cultivated only from Mexico to the present border of Costa Rica and Panama. This practice was reported first in 1530 by Peter Mártir (1944), possibly based on oral information supplied by one of the discoverers of Nicaragua. The best description, however, was written by Fernández de Oviedo, who lived in Nicaragua around 1520 (Fernández de Oviedo, 1959). His description of the shade trees is so clear that its identification is easy: *Gliricidia sepium*. There is a well known woodcut in Benzoni (1572) showing a cacao tree under the complete shade of a larger tree. Benzoni was in Nicaragua around 1550.

During the early part of the colonial period, cacao was planted on the Pacific slope of Mesoamerica, in areas of alternate dry and wet seasons. For some unknown reason, *Gliricidia sepium* was replaced by *Erythrina* during the seventeenth century. When cacao production moved to the Atlantic part of Central America, a region with permanent humidity, other species including *Inga* had to be used for shade.

In South America, the situation was different. Cacao grew wild in many places but its culture was unknown. The Indians chewed the pulp that surrounds the seed, and discarded the seeds. Numerous heaps of them were found in the forests by the Spaniards who recognized them as "the same that was cultivated in New Spain". Cacao cultivation was started with these seeds in several places in South America. More or less at the same time, seeds were introduced to Venezuela from Central America, and very likely with them, the cultural practices including the use of shade trees. If so, *Gliricidia sepium*, which grows wild in the lowlands of Venezuela, was the first tree to be used. Sometime in the last century, it was replaced by *Erythrina fusca* and *E. poeppigiana*. Around 1925, Pittier recommended abandoning *Erythrina* and replacing it with *Inga* (Pérez Arbelaez, 1937).

The problem of shade in cacao has focused on a wider approach than in coffee, including its economic and social aspects (Alvim, 1966). A modern technology was developed in Brazil, growing cacao without shade, with intense fertilization and weed control. However, prices and diseases have halted the spread of this new practice.

Several species of *Inga* are used to shade cacao, most of them in restricted areas. In Mexico, the commonest is *I. vera*, especially on the Pacific slope. In Guatemala, *I. micheliana*, *I. paterno* (= *I. jinicuil*), *I. sapindoides* and *I. vera* (de la Cerda, 1993). In Nicaragua, "*I. vera* commonly used by farmers, has the advantage of rapid development of a broad canopy which permits a non dense stand, in which trees can be planted at a distance of 9 × 9 m. It also shows more adaptability to poor soils. The extension of the canopy and the continuous recycle of the leaves determines a microclimate advantageous to cacao growth " (Thienhaus, 1993). In Costa Rica, the commonest *Inga* in the cacao plantations is *I. oerstediana*; *I. densiflora* and *I. vera* are planted on a small scale. *Inga edulis* is a very promising species; in CATIE, the cacao collection was planted in 1948–1952 under *I. edulis* and both cacao and shade trees are in very good condition. In Colombia, *Inga* are planted especially in the farms between 900–1300 m (Llano Gómez, 1947); in the lowlands *I. laurina* is sporadically planted and in Venezuela and the Antilles, the same species is used for shade; *I. punctata* is the most common *Inga* in the Venezuelan cacao farms (Martínez & Enriquez, 1981). In Ecuador, the more common species are *I. edulis* and *I. spectabilis* (Enriquez, 1996). In the Bahia region of Brazil, the most important cacao region in the Western Hemisphere, more than 125 species of native trees are found in the plantations. *Inga* is the most widely used, after *Spondias mombin* (CEPLAC, 1965). The three species of *Inga* are inga-cipo, *I. affinis* (= *I. vera*); ingá-mirim, *I. marginata*, and ingá-sabao, *I. nuda* (= *I. striata*) (Vinha & Silva, 1982). In the Antilles *I. vera* is found in Jamaica (Wright, 1949) and the Dominican Republic. In some of the Lesser Antilles, *I. fagifolia* (= *I. laurina*) is planted for shade and as a windbreak.

INGA AS SHADE IN COCA AND TEA CULTIVATION

According to Rostworowski de Diez Canseco (1981), *Inga* trees were planted as shade in the coca fields before the arrival of the Spaniards in Peru. Neither Garcilaso de la Vega (1960), who was a coca planter in the sixteenth century, nor Cobo (1956) who wrote a detailed account of coca production in the following century, refer to this practice. At present, *Inga* trees are planted as shade in some coca plantations, especially in northern Peru. According to Rusby (Cook, 1901), the alkaloid content of the leaves under shade is inferior to the leaves under full sun, but the leaves are of a more attractive appearance. The coca is often planted on sloping ground, with the soil free of weeds, and therefore exposed to erosion. In Bolivia, the species that is more common is *I. luschnathiana* (= *I. subnuda*), locally called ziquile (Cárdenas, 1969). In the mountains of northern Peru, the Trujillo coca (*Erythroxylon novo-granatense*) is planted under the shade of pacae, *I. feuillei* (Plowman, 1984).

Tea is planted in the eastern slopes of the Andes, especially in Peru, using as shade trees *Inga feuillei*, *I. fagifolia* (= *I. laurina*), *I. vera* and others (Carrasco, 1971). *Inga adenophylla* is extensively used for shade over tea in the Quillabamba valley, Cuzco, Peru between 1000–2000 m altitude (Pennington, 1997).

REFERENCES

Aguilar, M. 1845. Memoria sobre el cultivo del café, arreglada a la práctica que se observa en Costa Rica. Imprenta de la Paz, Guatemala.

Ackerman, A. & van Baer, P. 1992. El café orgánico: la sostenibilidad de un grano de oro. CDR Universidad Libre, Amsterdam.

Alvarado, J.A. 1935. Tratado de cafecultura práctica. Tipografía Nacional, Guatemala.

Alvim, P. de T. 1966. Problema de sombreamiento do cacaueiro. Cacau Atual. 3: 2–5.

Anon. 1958. Manual del cafetero colombiano. Federación Nacional de Cafeteros, Bogota.

Balmaseda, F.J. 1890. Tesoro del agricultor cubano. La Propagandistica Literaria, Habana.

Baralt, R. M. & Diaz, R. 1939. Resumen de la historia de Venezuela. Desclée-De Bower, Brujas, Belgica.

Beer, J. 1987. Advantages and disadvantages and desirable characteristics of shade trees for coffee, cacao and tea. Agroforest. Systems 5: 3–13.

Benzoni, G. 1572. La historia del Mundo Nuovo. P. & F. Tini, Venice: (facsimile edition, 1962, Akademische Druck, Graz).

Boyce, J.K., Fernández González, A., Furt, E. & Segura Bonilla, O. 1994. Café y desarrollo sostenible: del cultivo agroquímico a la producción orgánica en Costa Rica. Costa Rica-EFUNA, Heredia.

Bradford Burns, E. 1986. In Guatemala, 1875. University of California Press, Berkeley.

Budowski, G., Kass, D.C.L. & Russo, R.O. 1984. Leguminous trees for shade. Pesq. Agropecu. Brasil. 19: 205–222.

Cárdenas, M. 1969. Manual de plantas económicas de Bolivia. Imprenta ICTHUS, Cochabamba, Bolivia.

Carrasco, F. 1971. El defoliador del pacae, *Automolis inexpectata* Rothschild (Lepidoptera, Anctiidae) en el departamento del Cuzco. Revista Entomol. (Peru) 14: 140–142.

Castro, Y.G.P. de. 1952. Experiências sôbre germinaçao de sementes de *Inga striata*. Mimeograph, Forestry Service, Sao Paulo.

CEPLAC. 1965. Relatorio anual 1964. CEPLAC, Bahia, Brazil.

Chevalier, A. 1929. Les caféiers du monde. I. Lechevalier, Paris.

Cobo, B. 1956. Historia del nuevo mundo 1. Real Academia Española, Biblioteca de Autores Españoles, Madrid.

Cook, O.F. 1901. Shade in coffee culture. U.S.D.A. Div. Bot. Bull. 25.

Cowgill, W.H. 1954. Try growing coffee in sunhedges for quicker and greater production. Instituto Agropecuario Nacional, Guatemala.

Cramer, P.J.S. 1957. A review of literature of coffee research in Indonesia. CATIE, Turrialba, Costa Rica.

de la Cerda, C. 1993. Sombra y cultivos asociados al cacao en Guatemala. In: W. Phillips-Mora (ed.), Sombras y cultivos asociados con cacao: 181–182. CATIE, Turrialba, Costa Rica.

Dumont, A.B.C. 1833. Consideraciones sobre el cultivo del café en esta isla. Habana: Imprenta Fraternal (reprinted in Costa Rica (1835), San José, Imprenta de la Paz; 1990 Documentos Históricos, Imprenta Nacional, 35–75).

Ely, R.T. 1962. La economía cubana entre las dos Isabeles. 1492–1832. Aedita, Bogotá.

Enriquez, G.A. 1996. Sistemas de cultivo del cacao en el Ecuador. Mini-simposio sobre sistemas agroforestales, Bahia.

Escalante, E.E., Aguilar, A. & Lugo, R. 1987. Identificación, evaluación y distribución de especies utilizadas como sombra en sistemas tradicionales de café (*Coffea arabica*) en dos zonas del estado de Trujillo. Venezuela Forestal 3: 50–62.

Fernández de Oviedo, G. 1959. Historia general y moral de las Indias. I. Real Academia Española, Biblioteca de Autores Españoles, Madrid.

Fournier, L.A. 1988. El cultivo del cafeto (*Coffea arabica*) al sol o a la sombra: un enfoque agronómico y ecofisiológico. Agron. Costarric. 12: 131–146.

Franco, C.M. 1957. Sombramiento. 1 Curso de cafei-cultura, Campinas, Brasil: Instituto Agronómico, 161–166.

Fukunaga, E.T. 1957. Report on coffee cultivation in Peru. IICA, Report No. 53. Turrialba, Costa Rica.

Garcilaso de la Vega, I. 1960. Comentarios reales de los Incas. Universidad Nacional de Cuzco, Cuzco.

Guiscafré-Arrillaga, J. 1942. Effect of solar radiation intensity on the vegetative growth and yield of coffee. J. Agric. Univ. Puerto Rico 26: 73–90.

Kimenia, J.K. & Njoroge, J.M. 1988. Effect of shade on coffee: a review. Kenya Coffee 53: 387–391.

Lawrence, A. 1995. Farmer knowledge and the use of *Inga* species. In: D.O. Evans & L.T. Szott (eds.), Nitrogen fixing trees for acid soils: 142–151. Nitrogen Fixing Tree Association, Morrilton, Arkansas.

Lawrence, A., Pennington, T.D., Hands, M.R. & Zúniga, R.A. 1995. *Inga* species. In: D.O. Evans & L.T. Szott (eds.), Nitrogen fixing trees for acid soils: 130–141. Nitrogen Fixing Trees Association, Morrilton, Arkansas.

Lawrence, A. & Zúniga, R.A. 1996. The role of farmers' knowledge in agroforestry development: a case study from Honduras and El Salvador. The University of Reading, AERDD Working Paper 96/5.

Lecompte, H. 1899. Le café: culture, manipulation, production. Carré & Naud, Paris.

León, J. 1966. Central American and West Indies species of *Inga*. Ann. Missouri Bot. Gard. 53: 265–359.

Llano Gomez, E. 1947. Cultivo del cacao. Ministerio de Economía Nacional. Bogotá.

Lock, C.G.W. 1888. Coffee: its culture and commerce in all countries. London: E. & F.N. Spon, London.

Marrero, J. 1954. Especies del género *Inga* usadas como sombra del café en Puerto Rico. Caribbean Forest. 15: 54–71.

Martínez, A. & Enríquez, G. 1981. La sombra para el cacao. CATIE, Turrialba, Costa Rica.

Mártir, P. 1944. Décadas del nuevo mundo. Bajel, Buenos Aires.

Meléndez, G., Szott, L.T. & Ricse, A. 1995. Mineralización de nitrógeno de material foliar de *Inga*. In: D.O. Evans & L.T. Szott (eds.), Nitrogen fixing trees for acid soils: 35–41. Nitrogen Fixing Tree Association, Morrilton, Arkansas.

Montealegre, M. 1938. Estudios sobre el café. De la sombre. Revista Inst. Defensa Cafe 6: 359–372.

Mora, F. 1910. Colección de artículos publicados en La República en el debate sobre la industria cafetera. Tipografía Nacional, San José, Costa Rica.

Moral, P. 1955. La culture du café en Haiti: des plantations coloniales aux "jardins" actuels. Cahiers d'Outremer 31: 233–256.

Ospina, F. 1952. Cultivo del café. Nociones elementales al alcance de todos los labradores. In: J.M. Restrepo, (1952), Memorias sobre el cultivo del café: 51–74. Banco de la Republica, Bogotá.

Ostendorf, F.W. 1962. Review article: the coffee shade problem. Trop. Abstr. 17: 577–586.

Oviedo, G.F. de, see Fernandez de Oviedo, G.

Palm, C.A. & Sanchez, P.A. 1990. Decomposition and nutrient release patterns of the leaves of three tropical legumes. Biotropica 22: 330–338.

Patiño, V.M. 1969. Plantas cultivadas y animales domesticos en América Equinoccial IV. Plantas introducidas. Imprenta Departamental, Cali, Colombia.

Pennington, T.D. 1997. The Genus *Inga* – Botany. Royal Botanic Gardens, Kew.

Pérez Arbeláez, E. 1937. Manual del cacaotero venezolano. Cooperativa de Artes Gráficas, Caracas.

Pérez de la Riva, F. 1944. El café: historia de su cultivo y explotación en Cuba. Jesús Montero, La Habana.

Pérez Solano, V.M. 1983. Treinta y dos años de investigación sistemática y transferencia tecnológica del cultivo del café en Costa Rica 1950–1982. Oficina del Café, San José, Costa Rica.

Pérez Zeledón, P. 1910. Colección de artículos sobre política agrícola. Tipografía Nacional, San José, Costa Rica.

Perfecto, I., Rice, R.A., Greenberg, R. & Van der Voort, M.A. 1996. Shade coffee: a disappearing refuge for bio-diversity. BioScience 46: 598–608.

Phillips-Mora, W. (ed.) 1993. Sombras y cultivos asociados con cacao. CATIE, Turrialba, Costa Rica.

Pittier, H. 1929. The Middle American species of *Inga*. J. Dept. Agric. Porto Rico 13: 112–177.

Plowman, T. 1984. The ethnobotany of coca. In: G.T. Prance & J.A. Kallunki, (eds.), Ethnobotany in the tropics 1: 62–111. New York Botanical Garden, New York.

Preuss, P. 1901. Expedition nach Central und Sudamerika 1899–1900. Kolonial-Wirtschaflichen Komitees, Berlin.

Ridler, N.B. 1982. Implications of the new coffee technology in Central America. Desarrollo Rural de las Américas 14: 63–71.

Roskoski, J.P. 1981. Nodulation and N_2 fixation by *Inga jinicuil*. A woody legume in coffee plantations. Pl. & Soil 59: 201–206.

Rostworowsky de Díez Canseco, M. 1981 Recursos naturales renovables y pesca. Siglos XVI, XVII. Instituto de Estudios Peruanos, Lima.

Sáenz, N. 1895. Memoria sobre el cultivo del cafeto. J.J. Perez, Bogotá.

Semler, H. 1892–1900. Die tropische agricultur. Wismar, 4 vol. (El café. Revista Inst. Defensa Café 16: 110–116. 1945).

Thienhaus, S. 1993. Evaluación de diferentes leguminosas arbóreas como sombra del cacao en El Recreo, Nicaragua. In: W. Phillips-Mora (ed.), Sombras y cultivos asociados con cacao: p. 191. CATIE, Turrialba, Costa Rica.

Uribe, L. 1945. Arboles de sombrio en los cafetales en Colombia. Caribbean Forest. 6: 82–85.

Van der Laat, J. 1910. La sombra en los cafetales. Tipografía Nacional, San José, Costa Rica.

Vinha, S.G. & Silva, L.A.M. 1982. Arvores aproveitadas como sombreadoras de cacaueiros no sul de Bahia e norte do Espirito Santo. CEPLAC, Ilheus, Brasil.

Wellman, F.L. 1961. Coffee. Botany, cultivation and utilization. Leonard Hill, London.

Willey, R.W. 1975. The use of shade in coffee, cacao and tea. Hort. Abstr. 45: 791–798.

Wright, J. 1949. Shade and cocoa. Jamaica Department of Agriculture Extension Circular 28.

CHAPTER 8. *INGA* AND INSECTS: THE POTENTIAL FOR MANAGEMENT IN AGROFORESTRY

ILSE L. ACKERMAN, ELLEN L. McCALLIE
& ERICK C.M. FERNANDES

INTRODUCTION

The genus *Inga* is widely exploited in the neotropics for its edible fruit, for its fuelwood, and as a shade tree for coffee. As the genus *Inga* is known by neotropical farmers to attract a high diversity and abundance of insects to its vicinity, *Inga*'s entomology is of interest and importance to agriculturalists and ecologists alike. Identifying and understanding the interactions among the insects that seek out *Inga* for food or shelter is important for the development of appropriate agroecosystem management tools to take advantage of potential mutualistic or symbiotic associations involving *Inga* and insects. This chapter, divided into two parts, reviews the existing literature on the incidence and impact of insects on *Inga*. The first half of the chapter addresses *Inga*'s leaves, fruits, wood, and nectar as resources for a wide range of insects, as well as the functions insects serve *Inga* in pollination and defence. The second half of the chapter examines the ecological impact of insect-*Inga* associations and how this feature of *Inga* has been exploited in agroforestry system design with apparent success.

ASSOCIATIONS OF INSECTS AND *INGA*

Leaves of Inga as a source of food and shelter for insects

Grasshoppers, katydids, and larvae of Lepidoptera are among the insects feeding on *Inga* leaves for food (Koptur, 1983b). At least 17 families of lepidopteran larvae feed on *Inga*; the most prevalent are the families *Pieridae*, *Hesperiidae*, *Lycaenidae*, *Nymphalidae*, *Megalopigidae*, *Saturniidae*, *Notodontidae*, and *Geometridae* (see Table 1), (Koptur, 1983a, 1985; Callaghan, 1988). In a study by Koptur (1984a), the greatest numbers of caterpillars occurred on *Inga* during the dry season, while the lowest occurred at the end of the wet season. Caterpillar loads also differed among species of *Inga* (Koptur, 1984a). Earlier lepidopteran instars usually consumed young leaves, while later instars consume both young and old leaves (Koptur, 1983b).

While the above references do not indicate if the lepidopteran eggs are oviposited directly on *Inga* leaves, Bauza (1991) reports that in Puerto Rico the eggs, larvae, and pupae of the butterfly *Dismorphia spio* Godart use young *Inga vera* and *Inga laurina* plants (10 cm to 2 m tall) as host and food plants. *D. spio* oviposits on mature leaves, close to the secondary veins. No eggs were observed on young, old, or rough-textured leaves, however.

Older *Inga* leaves are used by leaf-miners, which consume the leaf mesophyll, and by wasps and small flies, which oviposit in the leaflet lamina or inside the rhachis and petiole, causing galls (Koptur, 1983b). Older leaves are also used as shelter for various leaf binding insects and spiders. Leaf binding insects include skipper larvae, small orthopterans, and microlepidopterans. The lepidopteran leaf binders scrape the upper epidermis and mesophyll of the leaf tissue, resulting in bare brown leaf sections (Koptur, 1983b).

Leaves of Inga as a fungal substrate for ants

Leaf cutter ants use the leaves of *Inga*, among those of many other plants, as a substrate on which to cultivate the fungus that is the sole food for their larvae. In Costa Rica, Nichols-Orians & Schultz (1990) observed *Inga* leaves being harvested by the leaf cutter ant species *Atta cephalotes*. In experiments relating leaf traits of *Inga* with susceptibility to harvesting by leaf cutter ants, physical and chemical characteristics of *Inga* leaves were determined to influence their suitability to cutter ants. Bioassays that required cutting before carrying showed that *A. cephalotes* preferred the younger, more supple leaves. Young *Inga* leaves are typically less tough, yet have higher levels of condensed tannins, than older leaves. When selecting on the basis of chemical cues alone, however, *A. cephalotes* always preferred mature leaves of *Inga oerstediana* Benth. to young leaves. As young leaves are determined to show greater inhibition of fungal pectinases, Nichols-Orians & Schultz (1990) suggest that *A. cephalotes* chooses leaves based on their suitability as a substrate for fungus cultivation. The authors' observations also suggest that *A. cephalotes* avoids harvesting from *I. oerstediana* when other plant species are readily available (Nichols-Orians & Schultz, 1990).

Further experimentation by Nichols-Orians (1991a) demonstrated that the levels of secondary chemical and nutritional components of *Inga oerstediana* Benth. leaves may be at least partially environmentally induced. In light-limited conditions, *Inga* seedlings had low leaf tannin concentrations, whereas in nutrient-limited conditions, leaf tannin concentrations were high. Despite the higher tannin concentrations, *A. cephalotes* preferred leaves of *Inga* seedlings grown in 20% light over those grown in 2% light. Nichols-Orians (1991b) suggests therefore that leaf selection by the leaf cutter ants may involve maximizing nutrient content while minimizing tannin content. Relative to other species, however, light-induced increases in condensed tannins in *I. oerstediana* may lower its susceptibility to leaf cutter ants and therefore increase the species' competitiveness relative to other species in treefall gaps (Nichols-Orians, 1991a).

The use of Inga fruit by insects

In Mexico, Hernandez-Ortiz & Perez-Alonso (1993) found *Anastrepha distincta* (*Diptera: Tephritidae*) to infest a small percentage of the pods of *Inga sapindoides* Willd. The fruit fly genus *Anastrepha* is one of the largest and most agriculturally damaging groups of insects in the American tropics and subtropics (Hernandez-Ortiz & Perez-Alonso, 1993). In a study by Malavasi & Morgante (1980), *Inga* was found to have the heaviest fruit fly load per fruit of the 14 most common fruit fly hosts in Brazil. An average of 4 larvae per fruit were found in the 88 *Inga* fruits surveyed. The vast majority of larvae were *Anastrepha*, although some *Silba* were also found. In general, fruits of non-domesticated plant species had a higher infestation index than semi-domesticated and domesticated species. For this the non-domesticated species are considered fruit fly "repositories", both for their high infestation levels as well as their wide geographic distributions. Wild fruit trees neighbouring commercial orchards may thus be of concern for farmers (Malavasi & Morgante, 1980).

Beetle larvae also commonly exploit the fruits of *Inga*. Several associations are reported between the genus *Conotrachelus* (*Coleoptera: Curculionidae: Molytinae*) and fruits of various *Inga* species. These are *Conotrachelus imbecilus* Fiedler with *Inga heterophylla*, *C. quadrinotatus* with *I. edulis*, and *C. costirostris*, *C. curvicostatus*, *C. geminus*, *C. incertus*, *C. inconcinnus*, *C. loripes*, *C. nitidiceps*, *C. persimilis*, and *C. quadrinotatus* with various species of *Inga* (Valente & Gorayeb, 1994). Valente & Gorayeb (1994) observed a synchrony between the larval development of these species and fruit maturity. Female *C. imbecilus* oviposit on fruit flesh in fully formed *Inga* pods. The changes between larval stages correspond to the ripening of the pods, such that when the pods senesce and fall to the ground, the larvae are in the fourth stage of development, ready to pupate in the soil.

Additional *Curculionidae* found on *Inga* include an adult of the subfamily *Brachyderinae*, observed on the leaves and trunk of *Inga* (McCallie, unpublished data) and the weevil *Neomastix numerus*, collected on *Inga* (Clark, 1993).

Larvae of the beetle families *Nitidulidae* ("sap beetles") and *Staphylinidae* ("rove beetles") were also found by Valente & Gorayeb (1994) to prey on *Inga* pods. Larvae of other insect orders found preying on the pods in this study were *Lepidoptera*, *Hymenoptera* and *Diptera* (*Lonchaeidae* and *Tephritidae*). The different orders displayed distinctive feeding patterns. Larvae of *Hymenoptera* and *Lepidoptera* were observed to prey on fruits during the fruit growth phase, larvae of *Diptera* and *Conotrachelus imbecilus* in the ripening phase, and larvae of the other beetle families (*Nitidulidae* and *Staphylinidae*) to prey during fruit senescence.

Different species of beetle larvae have their own characteristic effects on *Inga* seeds in the pod. Dipteran larvae only consume the aril of the seeds, turning it moist, brown, and easily removed from the membrane around the cotyledons. *Nitidulidae*, on the other hand, only feed on the cotyledons, creating galleries and reducing the cotyledons to fragments. In this case, the aril becomes brown, very dry, and easily removed. *Conotrachelus imbecilus*, until its third instar, feeds on *both* the arils and the cotyledons, making galleries as well. The aril's colour does not change, nor is it easily removed. In its fourth instar, both aril and cotyledons are destroyed. Valente & Gorayeb (1994) did not find a significant relationship between the frequency of larval infestations and *Inga* pod size. While the beetles *Nitidulidae* and *C. imbecilus*, or larvae of *Diptera* and the beetle *C. imbecilus* can be found within the same pod, the larvae have not been observed preying on the same seed.

Wood boring insects and Inga

The small (4–6 mm) wood-boring beetle, *Platypus ratzeburgi*, attacks at least two *Inga* species in Puerto Rico, *Inga vera* and *Inga fagifolia* (= *I. laurina*). It has also heavily infested a coffee plantation cultivated under *I. vera* shade (Gallardo, 1987).

Insects visiting Inga flowers

Although a thorough investigation of floral visitors and their impact on *Inga* has yet to be completed, Koptur (1983a), McCallie & Pardo (unpublished data), and Kühne & Uhle (unpublished data) provide preliminary observations. In montane cloud forest in Costa Rica, Koptur (1983a) reports floral visitors from 19 insect families, representing 5 orders: *Hemiptera*, *Coleoptera*, *Diptera*, *Hymenoptera*, and *Lepidoptera* (Table 1). By evaluating visitor size and visitor behaviour on and among flowers, Koptur (1983a) suggests visitor effectiveness as a pollinator. In addition to hummingbirds, species from the lepidopteran families *Hesperiidae*, *Sphingidae*, and *Uranidae* are cited as most likely to be effective pollinators.

In agroforestry systems near Manaus, Brazil, *Inga edulis* floral visitors included small to large wasps, small black bees, bumblebees (often with pollen on the head and/or proboscis), *Sphingidae* moths, small moths, and a myriad of tiny flying insects (McCallie & Pardo, unpublished data). *Sphingidae* moths were of the following species: *Enyo ocypete*, *Erinnyis ello ello*, *Manduca diffissa tropicalis*, *Xylophanes chiron nechus* and *Xylophanes loelia*. McCallie & Pardo (unpublished data) observed several wasp species chewing through the sepals and calyces of *I. edulis* buds and others piercing the corolla lobes. While thrips were commonly found on all parts of *I. edulis* blossoms, several types of eruciform larvae (typical of *Lepidoptera*, *Mecoptera*, and some *Hymenoptera*) were seen to feed on or near the floral nectaries and ovary. Kühne & Uhle (unpublished data) reported diurnal floral visitors to several unidentified *Inga* species near Tena, Ecuador to be primarily butterflies, bees and tiny flying insects. Descriptions of insects encountered on various species of *Inga* in Ecuador are provided in the appendix (Rodriguez & Onore, data from unpublished M.Sc. thesis).

Inga extrafloral nectaries and their visitors

Many of *Inga*'s more than 250 species (Mabberley, 1997) have extrafloral nectaries — plant glands located outside the flowers. In *Inga*, these nectaries are most commonly single, circular glands on the leaf rhachis between each pair of leaflets (Pennington, 1997). The secretion of nectar occurs as the leaf unfolds, and continues through its mature and expanded state. Since new leaves are produced year-round, extrafloral nectar is almost always available (Koptur, 1984a).

Nieuwenhuis von Uexkull (1907) in Jolivet (1996) was the first to speculate on a mutualistic relationship between plants and ants mediated by extrafloral nectaries. Since then, authors have interpreted the relationship as either mutualistic or symbiotic, and much evidence has accumulated to demonstrate that in most plants with extrafloral nectaries insects protect the plants from predation.

The role and composition of extrafloral nectar

Extrafloral nectar is presumed to be the primary attractant/reward for ant presence on *Inga*. Although several species of ants are often found on a single species of *Inga*, only one species of ant is usually hosted by any individual tree. There is an exception on large trees, where several ant species may exploit different regions of the tree. The species of ant on a tree or region of a tree typically corresponds to the nearest ant nest. While *Inga* does not produce any special domatia as shelter for the ants, occasionally ant nests may be found in the tree branches themselves, as in the case of *Dolichoderus bispinosa*, or in dead twigs (Koptur, 1984a).

TABLE 1. Insect taxa observed visiting *Inga* (adapted from Koptur, 1983a). Species are unidentified unless given.

Order	Family	Species
Hemiptera	Lygaeidae	
Coleoptera	Brentidae	
	Cerambycidae	
	Scarabidae	
Diptera	Bibionidae	
Hymenoptera	Euglossinae	*Eulaemma* sp., unidentified spp.
	Pompilidae	
	Vespidae	
Lepidoptera	Arctiidae	
	Ctenuchidae	*Ichoria quadrigutta, Cyanopepla scintillans* Butler
	Geometridae	*Microgonia* spp., unidentified sp
	Hesperiidae	*Astraptes anaphus annetta* Evans, *A. fulgerator azul* Reakirt; *A. galesus cassius* Evans, *Ouleus cyrna* Mab., unidentified spp.
	Ithomiidae	
	Noctuidae	*Mocis* nr. *repanda*, unidentified spp.
	Pericopidae	*Mesenochroa rogersi* Druce, undentified spp.
	Pieridae	*Actinote leucomelas* Bates, *Dismorphia crisia lubina, D. eunoe desine* Hewitson
	Pyralidae	*Herpetogrammas* sp., unidentified spp.
	Sphingidae	*Aelopos titan, Agrius cingulatus, Pachygonia subhamata, Pachylia ficus, Perigonia lusca, Xylophanes chiron,* unidentified spp.
	Uranidae	*Coronidia leachii* Latr.

Davidson (1988) reported that ant gardens (nests) are found disproportionately on individuals of *Inga* compared to other genera, and suggests that extrafloral nectaries are a particularly abundant resource relative to food resources provided by other trees. Ants are observed on both young leaves with actively secreting nectaries and old leaves without; however, ant activity is generally higher on the younger leaves. Koptur (1984a) noted increased ant activity at midday and midnight on *Inga densiflora*.

Extrafloral nectar contains sugar and amino acids. In measurements on *Inga densiflora* and *Inga punctata*, nectar sugar concentrations range from 28.5%–40.2% and 30.8–47.2% respectively, over a 24-hour period, with sugar concentrations slightly higher during the late morning and early afternoon (Koptur, 1984a). Koptur also found sugar composition of extrafloral nectar to be more variable than floral nectar, even when controlling the ambient conditions (Koptur, 1994). Generally, sugar concentrations in extrafloral nectars are substantially greater than in floral nectars and are hexose-dominant (Koptur, 1994). *Inga* nectars are known to contain the amino acids alanine,

121

glycine, isoleucine, proline, tyrosine and valine. Asparagine and glutamic acid were detected but unconfirmed in nectars from two species (Koptur, 1994).

Comparing floral and extrafloral nectars within a species, Koptur (1994) suggests that differences may result in part from their respective selective pressures. Cysteine, an amino acid thought to be important to ants, is found in only three out of eight floral nectars but found in four out of five *extrafloral* nectars in which amino acids were detected (Koptur, 1994). Note that ants are commonly-reported visitors to extrafloral nectaries, but are not observed visiting floral nectaries.

The roles of the insect patrons of extrafloral nectaries

Many species of ants, while visiting the foliar nectaries of *Inga*, provide protection of both young and mature leaves against a variety of insect herbivores by predation or by disturbing the herbivores until they leave (Koptur, 1984a). In experimental studies, leaves from which ants were excluded sustained greater damage than control leaves on which ants were allowed. Koptur (1984a) reported mean leaf damage on *I. densiflora* at <12% for the control and at 37% for leaves with ants excluded. In *I. punctata,* mean damage was 13% and 20%, respectively.

Although all ants common on *Inga* remove at least small caterpillars, ant species differ in their overall effectiveness in removal (Koptur, 1984a). The ant *Camponotus femoratus*, feeding at foliar nectaries of *Inga marginata* and *Inga ruiziana*, is effective against smaller invading species, while the formic acid of *C. femoratus* deters even large vertebrates. Davidson (1988) found that monkeys which feed on plant fruits and ant broods are uncomfortable and even leave individuals of *Inga* that have large *C. femoratus* populations.

In experimental studies, Koptur (1984a) found *Pheidole biconstricta*, a moderately-sized (3.5 mm) ant which forages in groups, to be most effective at removing herbivores from two species of *Inga* in Costa Rica. Least effective were the larger (6 mm) workers of *Monacis bispinosa*. When close to their nests, *M. bispinosa* ants forage in groups. Their foraging is so focused that they are not distracted by other insects on the leaves. The ants *Camponotus substitutus* and *Crematogaster limata palans* remove herbivores at an intermediate level of effectiveness. The workers of these species are small in size and, although they forage in groups, their overall numbers on an individual tree are less than those of *P. biconstricta.* Koptur (1984a) concluded that the most effective protectors are ants of small to intermediate size. She did not test the large (and rarer) solitary foragers, such as *Ectatomma*, but suggests that they are probably also effective, though much less common and less predictable (Koptur, 1984a).

Typically small caterpillars are physically removed from the leaf surface or are bitten repeatedly by ants until they voluntarily drop from the leaf. Ants also use a biting strategy on larger herbivores, such as katydids, beetles and large caterpillars. Koptur (1984a) notes two instances of a group of *Pheidole* ants surrounding, biting and stinging a large caterpillar before carrying it away. Some insects seem to defer to, or at least avoid, ants. Beetles often move to leaflet-tips, out of the range of ant patrols, which typically cover only the basal three-fourths of leaflet surfaces and the petioles and rhachises of leaves. Leafhoppers, on the other hand, generally abandon the leaf entirely upon encountering an ant (Koptur, 1984a).

At Yurimaguas, in the Peruvian Amazon, Fernandes (unpublished data) observed the giant, arboreal-foraging but ground-nesting ant, *Paraponera clavata* (*Hymenoptera: Formicidae*) in burrows at the base of *Inga edulis*. Although *P. clavata* appears to rely heavily on extrafloral nectar (Gobbitz, 1988), a study of nest site selectivity by *P. clavata* showed no correlation between the presence of extrafloral nectaries and location of nest sites by the ant (Belk *et al.*, 1989). Wetterer (1994), however, described *P. clavata* workers attacking a column of foraging leaf cutter ants (*Atta cephalotes*) on a *Rinorea deflexiflora* tree at La Selva, Costa Rica. It appears that, even with minimal predation, *P. clavata* can provide trees with effective protection against herbivory by leaf-cutting ants.

Trade-offs may exist between *Inga* defence by ants and other forms of defence. Evidence for this theory is found in a study by Koptur (1985). She compared herbivory of *Inga* in lowlands (where nectar-drinking ants are more active) and uplands (where nectar-drinking ants are less active). At both elevations extrafloral nectaries were present and actively secreting nectar. The same numbers of lepidopteran larvae were found at the two elevations. At the upper elevation, however, the concentrations of leaf phenolics were higher, suggesting that *Inga* compensates for the reduced ant defence with higher levels of secondary chemicals. Adult wasp and fly parasitization of caterpillars from upland *Inga* was also significantly higher than in those reared from lowland *Inga*. Since a smaller percentage of nectar is consumed by ants at the higher elevations, it suggests that the greater parasitization of lepidopteran larvae at higher elevations results from the greater availability of nectar to attract nectar-drinking wasp and fly parasitoids (Koptur, 1985).

INGA AS A COMPONENT OF AGROFORESTRY SYSTEMS

Studies of indigenous agroforestry systems in the Amazon have shown that *Inga* is an indicator of acid soils (Salick, 1989), an important component of managed fallow systems (Boom, 1990), and a trap crop for edible caterpillar species (Ribeiro, 1989). Many *Inga* species are used by farmers as shade trees for perennial crops such as cacao, coffee, and tea (Pittier, 1929; León, 1966; Carrasco, 1971). Lawrence (1995) reported that the main reasons that farmers like *Inga* as a shade tree is because "the leaves are a good fertilizer, the shade is perennial, the litter and shade provides good weed control, and the shade keeps the soil humid." On the basis of the widespread utilization of *Inga* by farmers, researchers have begun to test the integration of *Inga* into a range of agroforestry practices for soil conservation and sustainable food and wood production (Szott, 1987; Staver, 1989; Palm & Sanchez, 1990; Fernandes *et al.*, 1993, 1994; Riley & Smyth, 1993).

Impact of the insects associated with Inga on farmers and their cropping systems

Although *Inga* species have significant potential to improve and maintain soil productivity, there has been little research on the potential positive or negative effects to farmers and their farming systems arising from the association of a large number of insect species with *Inga*. For example, the sting of the giant ant *Paraponera clavata*, a common visitor of extrafloral nectaries on *Inga edulis* in the Peruvian Amazon (Fernandes,

123

personal observation) is very painful and the venom has been shown to significantly inhibit the ATPase activity of submitochondrial particles and F1-ATPase from bovine heart mitochondria (Zaitseva *et al.*, 1995). Carrasco (1971) reported that a defoliating caterpillar on *Inga* used as a shade tree in tea caused considerable discomfort to the tea pickers. It is not known whether this is the same species of caterpillar (hairy, luminous green, about 3 inches in length at maturity) that is found on the underside of *I. edulis* leaves in the Peruvian Amazon. Skin contact with the hairs of this caterpillar is excruciatingly painful (Fernandes, personal experience). Other reports include the introduction of a wood boring beetle of *Inga* to a coffee plantation (Gallardo, 1987). Given that *Inga* fruit have been shown to be hosts for fruit fly larvae, the presence of mature, fruit bearing *Inga* trees in fruit producing agroforestry systems is likely to cause increased predation and loss of the associated commercial fruit species. It should still be possible, however, to realize the soil conservation and wood production benefits of *Inga* in such systems if the *Inga* trees are regularly pruned for mulch, preventing flowering and fruit set. We did not encounter any published information about the possibility that insect visitors to the extrafloral nectaries are also fruit parasites.

Integrated pest management with Inga as a component of agroforestry systems
The attraction of a variety of ant, fly, wasp and spider species to the extrafloral nectaries of *Inga* species offers the potential to use these beneficial organisms to control insects that are detrimental to crops. An example of such an integrated pest management approach is reported by Matos *et al.* (in press). In this study, an agrosilvopastoral system was designed and established in 1991 to test the potential of *Inga edulis* as a nurse tree for mahogany (*Swietenia macrophylla*). Neotropical plantings of *Swietenia* have failed due to constant attacks of a lepidopteran pest, the mahogany shoot borer *Hypsipyla grandella* Zell. Repeated attacks by the shoot borer produce a witch's broom effect that adversely affects the commercial quality of the timber and can kill the young tree. Fernandes & Matos (1995) hypothesized that interplanting mahogany between two rows of *Inga* would provide several advantages that could reduce shoot borer attacks because of the following three factors:

A. *Inga* has a dense canopy that would physically shield the mahogany and make it difficult for *H. grandella* to locate its host.

B. *Inga* attracts a variety of insect species. Leston (1973) in (Koptur, 1983b) suggests that shade trees that support an "ant mosaic" in commercial plantations may increase protection of the crop plant by natural enemies. A species of *Hymenoptera* has been observed feeding on the extrafloral nectar. *Hymenoptera* are also reported to be important in the biological control of *H. grandella* (Yamazaki *et al.*, 1990). Other insects attract predators such as spiders (Fig. 1) which could prey on *H. grandella* as the moth tried to cross the "*Inga* gauntlet" to get to the mahogany. Recently, Greenberg *et al.* (1997) reported higher diversities of bird species in shade coffee systems involving *Inga* as the shade tree than in coffee systems with *Gliricidia* shade trees. It is

FIG. 1. Photo on the left shows mahogany plants planted without the *Inga* and *Schizolobium* nurse trees severely attacked by the shoot borer *Hypsipyla grandella* in the second year of growth. Photo on the right shows a large (3 inches in length) spider that was frequently seen on the *Inga* nurse trees in the mahogany-*Inga-Schizolobium* tree strips. (Photos: E.C.M. Fernandes).

likely that the insect associations of *Inga* have a major role in attracting these bird species. More birds in close proximity to the mahogany could also help keep *H. grandella* populations low.

C. The ant species that frequent the extrafloral nectaries are known to defend their territories against other insects (Koptur, 1984a; Wetterer, 1994) and could prevent *H. grandella* from reaching the mahogany.

The design involved three tree species: mahogany, *Inga edulis* and *Schizolobium amazonicum* with two rows of *Inga* (6 m apart) and a row of mahogany and *Schizolobium*. The idea revolved around physical lateral protection of the mahogany with dense canopies of *Inga* loaded with extrafloral nectaries and their insect patrons; and overhead protection by a sparse canopy of *Schizolobium*. The dense side shade and dappled overhead shade was seen as a means to force quick upward growth with a minimum of lateral branching in mahogany. As the trees grew, the lower branches of *Inga* were periodically pruned and the leaves added as a mulch at the base of the mahogany. The spatial and temporal associations of the plant species and the development of the *"Inga-Schizolobium* tunnel" are depicted in Fig. 2.

Results after four years of growth showed that the *Inga* nurse trees retarded the onset of the attack on interplanted mahogany by *H. grandella*. Although this trial did not have an open-planted mahogany control, adjacent mahogany plots growing without the protection of *Inga* were attacked by *H. grandella* in the second year after field establishment, at heights of around 2 m. In the *Inga*-mahogany association, however, the attacks took place largely in the third and fourth years and at heights of 3 to 7 m. There were two nutrient input levels in this experiment: low nutrient inputs (20 kg of P/ha) and a medium level of inputs (1 ton lime + 50 kg N, 20 kg P, 70 kg K). The nutrients were applied once to the trees and associated crops at the beginning of the experiment. After four years, the *Inga* canopy in the higher input system was considerably denser than the *Inga* canopy of the low input system (Fig. 3). Interestingly, 90% of the trees in the low input, sparse canopy *Inga*-mahogany system were attacked versus 70% in the high input, denser *Inga* canopy system (Fig. 4, Matos *et al.*, in press). The attack of *H. grandella* on mahogany increased once the mahogany trees grew taller than the *Inga* (6 m) as evidenced by the bushy growth and bifurcation of the mahogany stems (Fig. 4). The delayed attack by *H. grandella* on mahogany that was interplanted with *Inga* is significant

FIG. 2. Photo sequence 1991–1995 from top left to right showing the establishment of an agrosilvopastoral system with strips of mahogany intercropped with nurse trees of *Inga edulis* and *Schizolobium amazonicum*. Top row shows a tree strip in a crop sequence of upland rice followed by cowpea in year 1. Middle row (year 2 & 3) shows the cassava crop that followed cowpea and the *Inga*, mahogany and *Schizolobium* trees within each strip. Bottom row (year 4) shows an establishing pasture of *Desmodium ovalifolium* following the harvest of the cassava. The pasture is growing between two tree strips (*Inga*-mahogany-*Schizolobium*) spaced 20 m apart. The last photograph shows the structure of each strip, with a central row of mahogany and *Schizolobium* and two outer rows of *Inga edulis*. (Photos: E.C.M. Fernandes).

FIG. 3. The tree strip on the left received 20 kg P/ha while the strip on the right received 1 t lime, 50 kg N, 20 kg P and 70 kg K/ha during the cropping phase in year 1. The mahogany in the low input (20 kg P/ha) was attacked earlier and more severely than the mahogany in the high input plot. Most likely, the sparse *Inga* canopy in the low input treatment provided a much lower level of physical and biological protection than the dense *Inga* canopy in the high input treatment. (Photos: E.C.M. Fernandes).

Percentage of mahogany trees attacked by *Hypsipyla grandella* in low and high input agroforestry systems at Manaus, Brazil.

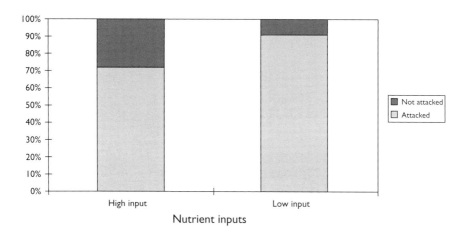

Total height, bifurcation height, and diameter at breast height (DBH) of mahogany in low and high input systems at Manaus, Brazil.

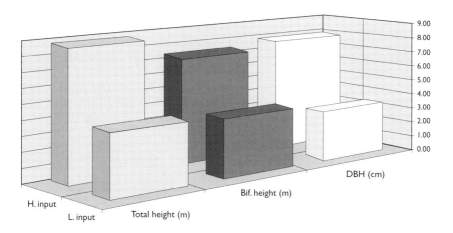

Fɪɢ 4. Top graph shows the percentage of mahogany trees attacked by *Hypsipyla grandella* in low and high input agroforestry systems involving *Inga edulis* nurse trees with mahogany at Manaus, Brazil. The bottom graph shows the total height, stem bifurcation height (point of attack by *H. grandella*) and diameter at breast height for the trees that were attacked in *Inga*-mahogany agroforestry systems at Manaus, Brazil (redrawn from Matos *et al.*, in press).

because the older trees were better able to survive the attack than the younger, open-grown mahogany trees. It is not possible to separate the effects of denser *Inga* canopies and higher potential insect predator populations and superior nutrition and faster growth of the mahogany trees. There is, however, a strong case for a controlled study of the pest-predator interactions and improved soil nutrient supply on reducing attacks of *H. grandella* on mahogany in an "*Inga-Schizolobium* tunnel."

CONCLUSIONS

A significant amount of information is available on the insect partners of the genus *Inga*, but relatively little has been done to apply this information in the management of agroecosystems. There appears to be considerable potential for harnessing the *Inga*-insect interactions for integrated pest management. The inclusion of *Inga* in an agroforestry system may reduce the need for pesticides and increase biodiversity. Research to further elucidate the many interactions between *Inga* and its various partners could help to optimize the productivity and diversity of agroforestry systems.

REFERENCES

Bauza, J.A.T. 1991. Biologia de *Dismorphia spio* (Godart) en Puerto Rico (*Lepidoptera: Pieridae: Dismorphiinae*). Caribbean J. Sci. 27: 35–45.

Belk, M.C., Black, H.L., Jorgensen, C.D., Hubbell, S.P. & Foster, R.B. 1989. Nest tree selectivity by the tropical ant, *Paraponera clavata*. Biotropica 21: 173–177.

Boom, B.M. 1990. Useful plants of the Parare Indians of the Venezuelan Guyana. In: G.T. Prance & M.J. Balick, (eds.), New directions in the study of plants and people. Advances Econ. Bot. 8: 57–76.

Callaghan, C.J. 1988. Notes on the biology of three *Riodinine* species: *Nymphidium lisimon attenuatum, Phaenochitonia sagaris satnius,* and *Metacharis ptolomaeus* (*Lycaenidae: Riodininae*). J. Res. Lepid. 27: 109–114.

Carrasco, F. 1971. El 'defoliador del pacae', *Automolis inexpectata* Rothschild (*Lepidoptera, Arctiidae*) en el Departamento del Cuzco. Revista Peruana Entomol. 14: 140–142.

Clark, W.E. 1993. The weevil genus *Neomastix* Dietz (*Coleoptera: Curculionidae, Anthonomini*). Coleopterists' Bull. 47: 1–19.

Davidson, D.W. 1988. Ecological Studies Of Neotropical Ant Gardens. Ecology 69: 1138–1152.

Fernandes, E.C.M, Davey, C.B. & Nelson, L.A. 1993. Alley cropping on an Ultisol in the Peruvian Amazon: Mulch, fertilizer and tree root pruning effects. In: J. Ragland & R. Lal, (eds.), Technologies for Sustainable Agriculture in the Tropics. Amer. Soc. Agron. Special Publ. 56: 77–96.

Fernandes, E.C.M., Garrity, D.P., Szott, L.A. & Palm, C.A. 1994. Use and potential of domesticated trees for soil improvement. Pp. 218–230. In: R.R.B. Leakey & A.C. Newton, (eds.), Tropical Trees: The Potential for Domestication. Proc. IUFRO Centennial Year (1892–1992) Conf. Edinburgh. HMSO, London.

Fernandes, E.C.M. & Matos, J.C. 1995. Agroforestry strategies for alleviating soil chemical constraints to food and fiber production in the Brazilian Amazon. Pp. 34–50. In: P.R. Seidl, O.R. Gottlieb & M.A.C. Kaplan, (eds.), Chemistry of the Amazon: Biodiversity, Natural Products and Environmental Issues. American Chemical Society, Washington, D.C.

Gallardo, C.F. 1987. *Platypus ratzeburgi* Chapuis (*Coleoptera: Platypodidae*): a new pest attacking coffee. J. Agric. Univ. Puerto Rico 11: 335–336.

Gilbert, L.E. & Smiley, J.T. 1978. Determinants of local diversity in phytophagous insects: host specialists in tropical environments. Symp. Roy. Entomol. Soc. London 9: 89–104.

Gobbitz, A. 1988. Ants that know when it is dinner time. Turrialba 38: 260.

Greenberg, R., Bichier, P., Cruz-Angon, A. & Reitsma, R. 1997. Bird populations in shade and sun coffee plantations in central Guatemala. Conservation Biol. 11: 448–459.

Hernandez-Ortiz, V. & Perez-Alonso, R. 1993. The natural host plants of *Anastrepha* (*Diptera: Tephritidae*) in a tropical rain forest of Mexico. Fla Entomol. 76: 447–460.

Jolivet, P. 1996. Ants and Plants: An Example of Coevolution. Backhuys Publishers, Leiden, The Netherlands.

Koptur, S. 1983a. Flowering phenology and floral biology of *Inga* (*Fabaceae: Mimosoideae*). Syst. Bot. 8: 354–368.

Koptur, S. 1983b. *Inga* (Guaba, Guajiniquil, Caite, Paterno). Pp. 259–261. In: D. H. Janzen, (ed.), Costa Rican Natural History. University of Chicago Press, Chicago.

Koptur, S. 1984a. Experimental evidence for defense of *Inga* (*Mimosoideae*) saplings by ants. Ecology 65: 1787–1793.

Koptur, S. 1984b. Outcrossing and pollinator limitation of fruit set: breeding systems of neotropical *Inga* trees (*Fabaceae: Mimosoideae*). Evolution 38: 1130–1143.

Koptur, S. 1985. Alternative defenses against herbivores in *Inga* (*Fabaceae: Mimosoideae*) over an elevational gradient. Ecology Publ. Ecol. Soc. Amer. 66: 1639–1650.

Koptur, S. 1994. Floral and extrafloral nectars of Costa Rican *Inga* trees: A comparison of their constituents and composition. Biotropica 26: 276–284.

Lawrence, A. 1995. Farmer knowledge and the use of *Inga* species. Pp. 142–151. In: D.O. Evans & L.T. Szott, (eds.), Nitrogen fixing trees for acid soils. Nitrogen Fixing Tree Research Reports (Special Issue). Winrock International and NFTA, Morrilton, AR, USA.

León, J. 1966. Central American and West Indian species of *Inga*. Ann. Missouri Bot. Gard. 53(3): 265–369.

Leston, D. 1973. The ant mosaic – tropical tree crops and the limiting of pests and diseases. Pest. Art. News Summ. (PANS) 19: 311–41.

Mabberley, D.J. 1997. The Plant Book: a Portable Dictionary of the Vascular Plants. Cambridge University Press, Cambridge.

Malavasi, A. & Morgante, J.S. 1980. Biology of 'fruit-flies' (*Diptera, Tephritidae*). II: Infestation indices in different food-plants and localities. Revista Brasil. Biol. 40: 17–24.

Matos, J.C. de S., Fernandes, E.C.M., Sousa, S.A.G., Perin, R. & Arcoverde, M.F. (In press). Performance of big-leaf mahogany in an agroforestry system in the western Amazon region of Brazil. Proceedings of an international meeting on "Big-Leaf Mahogany: Ecology and Management." San Juan, Puerto Rico, 22–24 October 1996. International Institute of Tropical Forestry, San Juan, Puerto Rico.

Nichols-Orians, C.M. 1991a. The effects of light on foliar chemistry, growth and susceptibility of seedlings of a canopy tree to an attine ant. Oecologia 86: 552–560.

Nichols-Orians, C.M. 1991b. Environmentally induced differences in plant traits: consequences for susceptibility to a leaf-cutter ant. Ecol. Publ. Ecol. Soc. Amer. 72: 1609–1623.

Nichols-Orians, C.M. & Schultz, J.C. 1990. Interactions among leaf toughness, chemistry, and harvesting by attine ants. Ecol. Entomol. 15: 311–320.

Nieuwenhuis von Uexkull, M. 1907. Extraflorale Zuckerausscheidung und Ameisenschutz. Ann. Jard. Bot. Buitenzorg 21: 195–327.

Palm, C.A. & Sanchez, P.A. 1990. Decomposition and nutrient release patterns of the leaves of three tropical legumes. Biotropica 22: 330–338.

Pennington, T.D. 1997. The genus *Inga* – Botany. The Royal Botanic Gardens, Kew.

Pittier, H. 1929. The middle American species of the genus *Inga*. J. Dep. Agric. Porto Rico 8: 117–177.

Ribeiro, B.G. 1989. Rainy seasons and constellations: The Desâna calendar. In: D.A. Posey & Balée, (eds.), Resource management in Amazonia: Indigenous and folk strategies. Advances Econ. Bot. 7: 97–114.

Riley, J. & Smyth, S. 1993. A study of alley-cropping data from northern Brazil. I. Distributional properties. Agroforest. Systems 22: 241–258.

Salick, J. 1989. Ecological basis of Amuesha agriculture, Peruvian Upper Amazon. In: D.A. Posey & Balée, (eds.), Resource management in Amazonia: Indigenous and folk strategies. Advances Econ. Bot. 7: 189–212.

Staver, C. 1989. Shortened bush fallow rotations with relay-cropped *Inga edulis* and *Desmodium ovalifolium* in wet central Amazonian Peru. Agroforest. Systems 8: 173–196.

Szott, L. 1987. Improving the productivity of shifting cultivation in the Amazon basin of Peru through the use of leguminous vegetation. Ph.D. Dissertation. North Carolina State University, Raleigh, USA.

Valente, R.D.M. & Gorayeb, I.D.S. 1994. Biology and description of immature forms of *Conotrachelus imbecilus* Fiedler (*Coleoptera: Curculionidae: Molytinae*) in *Inga heterophylla* fruits. Bol. Mus. Paraense Emilio Goeldi, N.S., Bot.

Wetterer, J.K. 1994. Attack by *Paraponera clavata* prevents herbivory by the leaf-cutting ant, *Atta cephalotes*. Biotropica 26: 462–465.

Yamazaki, S., Taketani, A., Fujita, K., Vasques, C.P. & Ikeda, T. 1990. Ecology of *Hypsipyla grandella* and its seasonal changes in population density in Peruvian Amazon forest. Japan Agric. Res. Quart. 24: 149–155.

Zaitseva, L.G., Zaitsev, V.G., Fenyuk, B.A., Pavlov, P.F., Vilenskaya, N.D., Ovchinnikova, T.V., Pluzhnikov, K.A., Grishin, E.V. & Grinkevich, V.A. 1995. Protein components of tropical ant venoms and their effect on mitochondrial H+-ATPase. Bioorganich. Khim. 21: 563–570.

APPENDIX

ENTOMOFAUNA OF THE GENUS *INGA* IN ECUADOR

GIOVANNA RODRIGUEZ & GIOVANNI ONORE

Pachylis laticornis (F.) (Hemiptera: Heteroptera: Coreidae)

Adult 30–36 mm long, colour metallic bronze with yellow nervations on hemielytra. The antenna with yellow-orange bands.

Colonies of adults and larvae suck the sap from the young branches. After a few days, these branches dry up. If the insects are disturbed while they are feeding, they emit a stinking liquid.

D'Aráujo e Silva *et al.* (1968) mention two other species of Coreidae (*Hypselonotus interruptus* Hahn., 1821 and *H. subterpunctatus* Am. & Serv., 1843) as *I.* spp. pests.

Host plants: *I. edulis, Schizolobium parahybum**

Poekilloptera sp. (Hemiptera: Homoptera: Flatidae)

Small plant hopper 22–26 mm long, flattened laterally; head, pronotum and the base of the wings orange; white wings with numerous black spots.

The juveniles are gregarious and can be found in the inferior part of branches, covered by abundant white wax secreted by themselves. The adults are always present in the colony of juveniles.

D'Aráujo e Silva *et al.* (1968) found *P. phalenoides* (L., 1758) as a pest of *I.* spp.

Host plants: *I. edulis* and *I. marginata**

Enchenopa albidorsa Fairm. (Hemiptera: Homoptera: Membracidae)

Body 8–11 mm long, black with one white and one orange dot.

The anterior part of the pronotum with spiny processes; orange legs. Adults and juveniles are found together in the apical part of branches from which they suck the sap.

Host plant: *I. edulis**

Heteronotus sp. (Hemiptera: Homoptera: Membracidae)

Treehopper 12–14 mm long; very brilliant body. Pronotum anteriorly with two thin spines and posteriorly with spherical processes.

It is found in mixed colonies of adults and juveniles on young leaves.

Host plant: *I. edulis**

---

* Observations from Province of Pichincha, locality Rio Pitzara, 00°15'N, 79°05'W. (See Chapter 2, Appendix 1 for full site description).

Umbonia spinosa (Houttuyn, 1766) (Hemiptera: Homoptera: Membracidae)

Adult 12–15 mm long, green with six reddish lines that join themselves in the apical part of the spiniform dorsum.

The female cuts the bark of the tree with her ovipositor and lays eggs on both sides of the cut; the female takes care of the juveniles which form large colonies on the apical part of branches. They suck the sap and the liquid excrement produced by the insects is colonized by black fungi.

D'Aráujo e Silva *et al.* (1968); Gara & Onore (1989) report similar damage to plants. D'Aráujo e Silva *et al.* (1968) also mention the Membracid *Umbonia curvispina* Stal, 1869 related to *I.* sp. meanwhile Martorell (1945) found *Monobelus fasciatus* (Fabricius) and *Nessorhinus gibberulus* Stal in *I. laurina* and *I. vera* respectively.

Host plant: *I. edulis**

Phoebis argante (F. 1775) (Lepidoptera: Pieridae)

Larva at final stage brown, 30–35 mm long; solitary and prefer to feed on young leaves. The pupa measures 27–32 mm, wider in the middle. Duration of pupal stage 8–10 days. The adult flies during the day from 10 am to 4 pm. The male has a wing span of 60–70 mm, the dorsal side of the wings yellow bordered with black. The female is a similar size to that of the male, but her wings are pale yellow with dispersed black dots.

De Vries (1987) and Martorell (1945) found the same species feeding on *I. vera*; D'Aráujo e Silva *et al.* (1968) mention it as a pest of *I. affinis, I. striata* and *I. uraguensis.*

The larva of this species was found feeding on the young leaves of *I. alata**

Syssphinx quadrilineata (Grote & Robinson, 1867) (Lepidoptera: Saturniidae: Ceratocampinae)

Eggs lentil shaped, of a very light green colour. Larva green with a crest of branched urticant reddish hairs on the thorax. Pupa approximately 50 mm long, covered with spiny processes. Adult with a wing span of 70–90 mm, anterior wings brown with two tranversal lines, posterior wings with a black dot surrounded by wine coloured stain. From pupa collected in the field, we found a parasite, Tachinidae (Diptera) that acts as a biological control.

D'Aráujo e Silva *et al.* (1968), report *Syssphinx molina* (Cramer, 1781) larvae, feeding on *I. sessilis.*

Larvae feed on mature leaves of *I. spectabilis**

Adeloneivaia jason (Boisduval, 1872) (Lepidoptera: Saturniidae: Hemileucinae)

Larva brown with two golden triangular stains on both sides. Pupa measures 45–50 mm, covered with spiny processes, the caudal part prolongs itself into a bidentate structure 5 mm long. Duration of pupal stage is 18–22 days. Adult with a wing span of 90–105 mm, orange; antenna plumose, turning filiform in the apical segments. Anterior wing with a white dot in the middle part and two brown tranversal lines, posterior wing uniformly orange.

A Tachinidae (Diptera) has been observed as a biological control in the pupal stage.

Larvae feed on *I. spectabilis**

Automeris fieldi Lemaire, 1969 (Lepidoptera: Saturniidae: Hemileucinae)

Eggs pearl white, 1.3 × 1.1 mm, which take 7–9 days to hatch. Larva of last stage green and it measures 75–85 mm; dorsal surface covered with urticant branched hairs; the entire larval stage lasts 80 to 88 days. Pupa oval, 35–42 mm long surrounded by the cocoon made of thick brown silk; pupal period 15–20 days. Adult with strong sexual dimorphism. Male, 65–70 mm of wing span; plumose antenna; dark brown thorax; orange abdomen; anterior wing beige; posterior wing orange with an eye-like spot surrounded by a golden ring. Female, 86–100 mm wing span; filiform antenna; dark brown thorax and orange abdomen; anterior wing wine coloured, posterior wing similar to that of the male. Gregarious larvae in the first stages, becoming solitary in the later ones.

A Tachinidae (Diptera) has been observed as a biological control in the pupal stage. The species was determined following Lemaire (1971).

The larvae feed on *I. alata*, *I. edulis*, *I. marginata*, *I. silanchensis* and *I. spectabilis**

Automeris banus argentifera Lemaire, 1966 (Lepidoptera: Saturnidae: Hemileucinae)

Larva green with urticant branched hairs. To begin pupation, the larvae stick together the leaves with silk and then knit a brown cocoon. Pupa oval, 35 mm long and wrapped in a cocoon of brown silk. Duration of pupal stage 15 days. Adult male has 70–85 mm wing span, plumose antenna; brown thorax; orange abdomen; anterior wing beige with some silver scales; posterior wing orange with a black eye-like dot. Adult female 85–115 mm wing span; filiform antenna; colour of wings similar to that of the male's but a bit darker.

Lemaire & Vénédictoff (1989), registered the moth but with no host plant.

Host plant: *I. marginata**

Automeris belti zaruma Shaus, 1921 (Lepidoptera: Saturniidae: Hemileucinae)

Green larva covered by urticant hairs. Oval pupa 3.4–4.5 mm long, surrounded by a cocoon of brown silk. Adult male with 70–80 mm of wing span, plumose antenna, brown thorax, orange-yellowish abdomen; anterior wing brown with an evident dark transversal line and the hind wing with a black spot. Adult female with 95–115 mm wing span; filiform antenna; colour similar to that of the male, although the wings are darker.

D'Aráujo e Silva *et al.* (1968), report the larva of *Automeris illustris* Walker, 1855 as a defoliator of *I.* spp.

Larvae feed on *I. edulis**

Hemiceras spp. (Lepidoptera: Notodontidae)

This genus contains at least five different species that feed on *Inga* spp. The mature larva measure 30–35 mm, cephalic capsule yellowish and body with longitudinal and transversal lines wine-red; gregarious and when disturbed they lift their head in a characteristic way, remaining in this position for some minutes; very active during the day, and can extensively defoliate *Inga* spp. Gallego (1949) describes similar damage to these plants. Duration of pupal stage 8–12 days; this takes place in the leaf litter on the floor or exceptionally on dead leaves that remain sticking to the branches of the tree.

The field observations indicate that larvae are attacked by fungi with white fructifications, possibly of the genus *Beauveria*.

Gallego (1949), describes *Hemiceras cadmia* on *I. edulis* and *I. densiflora*.

Host plants: *I. edulis, I. marginata* and *I. spectabilis**

Rosema sp. (Lepidoptera: Notodontidae)

Pupal stage lasts 8–12 days. Adult with wing span of 31–34 mm; body green with the anterior border of the pronotum brown; anterior wing dark green with dark apical border, posterior wing orange and grey.

Host plants: *I. edulis* and *I. spectabilis**

Lophocampa sp. (Lepidoptera: Arctiidae)

Last larval stage measures 30–35 mm with sulfur yellow hairs, cephalic capsule brown. Pupa 23–26 mm, enclosed in an oval yellow cocoon, 25–32 mm long, sticking on the underside of mature leaves. The adult emerges after 8–10 days of pupation. Chalcidoideae (Hymenoptera) acts as a biological control of the pupa.

Carrasco (1971) reports another Arctiidae, *Automolis inexpectata* Rothschild as a pest of *I. feuillei, I. laurina* and *I.* spp.

Host plants: *I. edulis, I. marginata* and *I. oerstediana**

Pelidnota soderstromi Ohaus, 1908 (Coleoptera: Scarabaeidae: Rutelinae)

Beetle of 25–28 mm long, colour metallic green-gold; feeds on flowers and young leaves.

Host plant: *I.* spp.

Rutela histrio Sahlber, 1823 (Coleoptera: Scarabaeidae: Rutelinae)

Beetle 11–13 mm long, colour yellow with some black spots. It is found eating flowers and young leaves.

D'Aráujo e Silva *et al.* (1968) found *Rutela lineola* (Linné, 1767) on *I.* spp. feeding on flowers and leaves.

Host plant: *I. edulis**

Macraspis melanaria Blanchard, 1908 (Coleoptera: Scarabaeidae: Rutelinae)

Adult measures 22–14 mm, colour piceus with a strongly developed scutellum. Chews flowers and leaves of *I. edulis**

Chrysophora chrysochlora Latr., 1811 (Coleoptera: Scarabaeidae: Rutelinae)

Very attractive adult, 30–34 mm long, colour metallic green with elytra heavily vermiculate and punctate, while pronotum and head are quite smooth; legs copper blue colour. Strong sexual dimorphism, the male with very developed metathoraxic tibia, while the females have the same structure in a normal size.

The Cofan indians of Ecuadorian Amazonia use the exoskeletons to make necklaces and other body ornaments. The "colonos" collect them and use them in substitution for Cantaridae beetles in the erroneus belief of getting sexual benefits.

Its distribution is typically Amazonian, but it has recently been found in Esmeraldas province on the western side of the Andes. This is evidence of a natural trans-Andean pass between Esmeraldas province and Sucumbios province through the Chota Valley.

According to Briceño (1989), a rutelid *Bolax palliatas* Burm., skeletonizes the leaves of *I.* sp.

They feed on young leaves of *I.* spp. on the sides of the rivers.

Deroconus sp. (Coleoptera: Curculionidae: Hypsonotini)

Weevil 15–22 mm long; black, with thin rows of punctures on the elytra. These insects are active during the night, gnawing the margins of young leaves. During the day, they hide under leaves and can cause serious damage.

Gara & Onore (1989), report that the same genus defoliates *I. edulis*.

Host plant: *I.* sp.

Naupactus sp. (Coleoptera: Curculinoidae: Naupactini)

Beetle, 12–15 mm long, colour black with lateral green stains. It feeds on young leaves, attacking the external margins and leaving characteristic teeth prints.

Host plant: *I. spectabilis**

Waldehemia atripennis (F.) (Hymenoptera: Symphyta)

Small wasp (13–14 mm wing span), colour black with reddish pronotum. Larva eruciform (caterpillar-like), colour yellow, active during the day on the upper side of the leaves; they feed on leaves and skeletonize them, and as a consequence, the leaves dry out and shrink. They can cause serious damage to the leaves of attacked plants. Similar damage is caused by other Symphyta like *Waldhemia pellucida* Konow, *Proselandria delicatula* (Kirby) and *Hemidianeura* sp.

Host plants: *I. edulis** and *I.* spp.

Atta cephalotes (L., 1758) (Hymenoptera: Formicidae)

Reddish leaf cutting ant. They attack the leaves and can completely defoliate trees. The attacks leave characteristic small teeth prints over the remnants of leaves. Similar damage is also caused by *Atta sexdens* (L., 1758).

Host plants: *I. edulis**, *I. spectabilis** and *I.* spp.

Anastrepha spp. (Diptera: Tephritidae)

Known commonly as fruit flies. Size similar to that of domestic fly, colour yellow; wings with dark stains. They attack the fruit pod. When the eggs hatch, the larva cause the fruit to rot.

Similar observations were made by D'Aráujo e Silva *et al.* (1968). Bruner *et al.* (1975) report *Ephestia neuricella* Zeller (Lepidoptera: Phycitidae) as a borer of *I. vera* seeds.

Host plants: *I. edulis**, *I. spectabilis** and *I.* spp.

REFERENCES (FOR APPENDIX)

Briceño, A.J. 1989. El escarabajo esquelitizador de las hojas de Guamo, *Bolax palliatus* Burm. (Mimeografiado).

Bruner, A.J., Scaramuzza, L.C. & Otero, A.R. 1975. Catalogo de los insectos que atacan a las plantas económicas de Cuba. Havana.

Carrasco, F. 1971. El "defoliador del Pacae", *Automolis inexpectata* Rothschild (*Lepidoptera: Arctiidae*), en el departamento del Cuzco. Revista Peruana Entomol. 14: 140–142

D'Aráujo e Silva, A.G., Rory Gonçalves, C., Monteiro Galváo, D., Lobo Gonçalves, J.A., Gomes, J., Do Nacimiento Silva, M., & De Simoni, L. 1968. Cuarto Catalogo dos insectos que vivem nas plantas do Brasil, seus parasitos e predadores. II, 1er & 2ndo tomo. Río do Janeiro, GB, Brasil.

De Vries, P.J. 1987. The Butterflies of Costa Rica and their Natural History. *Papilionidae, Pieridae, Nymphalidae*. Princeton University Press.

Gallego, L.M. 1949. Estudios entomológicos. I. Los gusanos de los Guamos. Revista Fac. Nac. Agron. Medellín Univ. Antioquia 10(34): 121–127.

Gara, R.I. & Onore, G. 1989. Entomología forestal. Proyecto DINAF-AID. Quito. Ecuador.

Lemaire, C. 1971. Révision du genre *Automeris* Hubner et des genres voisins (Lep. Attacidae) [1ére partie]. Mém. Mus. Nat. Hist. Natl. (N.S.), sér. A, Zool. 68: 1–232.

Lemaire, C. & Vénédictoff, N. 1989. Catalogue and Biogeography of the Lepidoptera of Ecuador. *Saturniidae*, with a description of a new species of *Meroleuca* Packard. Bull. Allyn Mus. 129: 1–60.

Martorell, L.F. 1945. A survey of the forest insects of Puerto Rico. Part 1. J. Agric. Univ. Puerto Rico 29(3): 69–608.

CHAPTER 9. SILVICULTURAL TRIALS OF MAHOGANY (*SWIETENIA MACROPHYLLA*) INTERPLANTED WITH TWO *INGA* SPECIES IN AMAZONIAN ECUADOR

DAVID A. NEILL & NIXON REVELO

INTRODUCTION

American mahogany (*Swietenia macrophylla* King and related species: *Meliaceae*) is one of the most coveted of tropical hardwoods and fetches high prices in international timber markets. *Swietenia macrophylla* is native to lowland tropical moist and dry forests throughout much of Central and South America, but has been intensively exploited throughout its range for more than four centuries and is now severely depleted in natural forest stands (Lamb, 1966; Gullison *et al.*, 1996). In order to supply the demand for mahogany timber, some attempts have been made to grow the trees in plantations, but many such trials have met with failure. One of the limiting factors for production of mahogany is that juvenile trees, in particular, are subject to infestation by a shoot borer, *Hypsipyla grandella* Zell. (*Lepidoptera: Pyralidae*) which consumes the apical meristem and young green shoots, resulting in desiccation of the shoots, production of new lateral shoots which in turn may be attacked, and sometimes in death of the juvenile tree. In the highly dense populations of mahogany in artificial plantations, the effect of the shoot borer is often very severe. Lamb (1966) suggested that mahogany plantations should be established with protective vegetative cover, such as in cleared lines in secondary forest; when mahogany is grown in mixed stands with other tree species, infestation of juvenile trees by the shoot borer may be much less severe. (See Ackerman *et al.*, this volume).

We report here on initial results of silvicultural trials of mahogany grown in combination with two *Inga* species, *I. edulis* Mart. and *I. ilta* T.D. Penn., at the Jatun Sacha Biological Station in Amazonian Ecuador. The trials include an experimental control of mahogany grown alone in pure stands. The present experiment was initiated as a follow-up to the *Inga* species trials carried out during 1992–1996 at Jatun Sacha and several other sites in the American tropics (reported by Pennington, this volume). We hypothesized that the *Inga* trees interplanted with mahogany, besides providing vegetative cover and protection from the mahogany shoot borer, would contribute to improved growth of mahogany through production of leaf litter and incorporation of organic matter and nitrogen in the soil. We report here the results of the initial 31 months (2 years 7 months) of growth of the *Swietenia* – *Inga* trials at Jatun Sacha.

SITE CHARACTERISTICS

Jatun Sacha Biological Station is located on the south bank of the upper Napo River in Napo province, Ecuador, about 40 km east of the base of the Andes, at 400 m elevation (01°04'S , 77°36'W). Mean annual rainfall is 3800 mm, distributed relatively evenly throughout the year; May and June are usually the wettest months and there are two short periods of relatively lower rainfall, in July–August and December–January, but in no month is the mean rainfall less than 100 mm (precipitation records at Jatun Sacha, 1988–1995). Mean annual temperature is about 24°C, with little annual variation. The site is transitional between Tropical Wet and Tropical Moist Life Zones (Holdridge, 1967). The biological station includes a forest reserve of 1800 ha, and also houses the Amazon Plant Conservation Center and the Ishpingo Botanical Garden which carry out programmes of applied research on conservation and utilization of plant species native to Amazonian Ecuador as well as agroforestry extension programmes with local farmers.

The *Swietenia – Inga* silvicultural trial was established on an abandoned pasture on property of the Cabañas Aliñahui tourist lodge, adjacent to Jatun Sacha Biological Station. The 3 ha site is nearly level with adequate drainage. Soil at the site is classified as Oxic Dystropept (USDA Soil Taxonomy system), highly acidic, low in nutrients and high in aluminum toxicity. The original forest cover was removed in the early 1970's, and the site was subjected to grazing by beef cattle for about 20 years, up until April 1994, a few months before establishment of the silvicultural trials. Vegetative cover in the pasture was almost exclusively the introduced African grass *Brachiaria decumbens* Stapf. Some woody plants were present, particularly near the edge of the pasture, including *Psidium guajava* L. and *Vernonia canescens* Kunth.

SPECIES CHARACTERISTICS

Swietenia macrophylla, the most widespread species in the genus, occurs on the mainland of Central and South America from Mexico south to Bolivia and Brazil, in lowland moist and dry tropical forests. Throughout most of its range, mahogany is found in scattered stands at low population densities, but in some areas such as the Yucatán Peninsula and Petén region of Mexico-Guatemala-Belize, it occurs at relatively high densities (Lamb, 1966). In Amazonian Ecuador, mahogany is quite rare and is known from only a few localities. One of the largest known concentrations in Ecuador was in the Arajuno River watershed, in primary forest about 10 km south of the study site, but that population was decimated by illegal logging activity in the early 1990's.

Inga edulis is a widespread species, occurring throughout tropical South America east of the Andes; it is widely cultivated in its natural range and also has been introduced to Central America (Pennington, 1997). *Inga ilta* is restricted in distribution to Amazonian Ecuador and northern Amazonian Peru. Both species produce edible fruits; *I. ilta* is one of only two *Inga* species for which the seed (embryo) is also edible and rich in protein (Pennington, 1997; Pennington & Robinson, this volume). Both species are cultivated by the local Quichua inhabitants of the upper Napo region and are common in secondary vegetation near the study site. Both *Inga* species produce abundant root nodules colonized by symbiotic nitrogen-fixing bacteria.

O = *Inga* species   X = *Swietenia macrophylla*

```
O O O O O O O O O O O
O X O X O X O X O X O
O O O O O O O O O O O
O X O X O X O X O X O
O O O O O O O O O O O
O X O X O X O X O X O
O O O O O O O O O O O
O X O X O X O X O X O
O O O O O O O O O O O
O X O X O X O X O X O
O O O O O O O O O O O
```

FIG. 1. Diagrammatic representation of a treatment plot of mahogany (*Swietenia macrophylla*) with *Inga* species. Mahogany trees were planted at 8 m × 8 m spacing, *Inga* trees at 4 m × 4 m spacing, such that the nearest neighbour of each mahogany was always an *Inga*. The plot contains 25 mahogany and 96 *Inga* trees.

METHODS

Seeds of *Swietenia macrophylla* were obtained from a single parent tree growing in mature forest about 10 km from the study site. Seeds of the two *Inga* species were obtained from several trees growing in secondary vegetation in the vicinity of the site. Seeds were germinated in planting beds, and seedlings were then transferred individually to 9" × 12" plastic bags filled with topsoil. When the plants attained a height of about 70 cm, they were considered ready for planting out; for the *Inga* species this was about 3 months following germination, and 5 months for mahogany.

A randomized block design was chosen for the *Swietenia – Inga* trials, with two treatment plots and one control plot per block, and three replicate blocks. In each plot, 25 mahogany trees were planted in a systematic array of 5 × 5 plants, with 8 m between plants. In each of the treatment plots, *Inga* trees of one species were planted between the mahoganies at 4 m spacing between plants. An extra row of *Inga* was planted around the perimeter of the plot. The nearest neighbour of each mahogany plant was therefore always an *Inga*, and the treatment plots contained 25 mahogany trees interplanted with 96 trees of either *Inga edulis* or *I. ilta* (Fig. 1). The control plots contained 25 plants of mahogany alone, planted at 8 m centres. The plantation trial contained a total of 225 mahogany plants and 288 plants of each *Inga* species.

Prior to planting, the pasture grass was removed with machetes for a radius of 1.5 m around each planting hole, and all woody vegetation within the plots was cut at ground level. The replicate blocks were adjacent to one another within the former pasture, and the site was nearly level, but some environmental variation among blocks was evident at the time of planting: soil compaction appeared to be more severe in Block 2, located in the centre of the former pasture, than in Blocks 1 and 3 which were located near the edges of the pasture. Woody vegetation was also more abundant at the edges of the former pasture than in the centre.

143

Soil samples, including presence of vesicular-arbuscular mycorrhizae and ectomycorrhizae, were obtained at the time of plantation establishment. The soil is to be resampled in each plot after three years of growth; soils data from this experiment will be reported in a subsequent paper.

The *Swietenia – Inga* trial was planted out in August 1994. Maintenance consisted of removal of the grass around each plant at 3 month intervals until the crowns of the juvenile plants were well above the level of the pasture grass (about 1.5 m). Woody vegetation regenerating spontaneously in the plots was removed at 6 month intervals, until crown closure of the *Inga* trees effectively eliminated competing vegetation. Measurements of the juvenile trees — height and diameter at breast height (DBH = 1.3 m) — were taken at one and two years from plantation establishment, together with information on mortality and incidence of attack of mahogany trees by the shoot borer. A third measurement was taken at 2 years 7 months (31 months: March 1997), the results of which are reported here.

RESULTS

Growth rates were highest for *Inga edulis*, which produced plants with broad open crowns and attained crown closure by 18 months from plantation establishment. At 2 years 7 months, *I. edulis* attained a mean height of 10.8 m and diameter of 13.8 cm (Table 1). The shade under the *I. edulis* trees was relatively light, and growth of herbaceous plants in the plots was suppressed but not entirely eliminated. Many of the *I. edulis* trees flowered and fruited at 2 years. Growth of *I. ilta* was somewhat slower, with a mean height of 7.6 m and diameter of 10.6 cm. The *I. ilta* trees produced more dense, compact crowns which cast a deeper shade than *I. edulis*; under *I. ilta* trees herbaceous plants were almost entirely eliminated. In Blocks 1 and 3, crown closure of *I. ilta* was obtained at about 2 years; the trees in Block 2 grew more slowly and just began to attain crown closure at 2 years 7 months. Because the leaflets of *I. ilta* are larger and more coriaceous than those of *I. edulis*, leaf litter produced by the former takes longer to decompose, and the plots with *I. ilta* were notable for the thick layer of litter beneath the trees.

Swietenia macrophylla plants grew as straight, unbranched poles except in cases where attack of the apical meristem by the shoot borer insect resulted in necrosis of the leader shoot and subsequent production of lateral branches. The crowns of mahogany trees grown with *Inga edulis* were, in general, partially shaded by the *Inga* after the first 18 months of growth.

Mean height for mahogany interplanted with *I. edulis* was 6.7 m; with *I. ilta* 6.3 m; and for the control plots of mahogany alone, 4.2 m. Diameter growth was greatest for mahogany interplanted with *I. ilta*, with a mean of 5.8 cm; interplanted with *I. edulis*, 5.1 cm; mahogany in control plots, 4.6 cm mean diameter (Table 1).

For the experiment as a whole, the observed differences among treatments in mahogany height and diameter were highly significant (two-way ANOVA, 2 df, $p < 0.001$). The differences in height and diameter among blocks (Table 2) was also highly significant (2 df, $p < 0.001$) as was the treatment × block interaction (4 df, $p < 0.001$).

TABLE 1. Height, diameter at breast height, and survival of *Swietenia macrophylla*, *Inga edulis*, and *Inga ilta* at 2 years 7 months from plantation establishment, at Jatun Sacha Biological Station, Ecuador. For height and diameter, the value on the first line is the mean; the values beneath (in parentheses) are upper and lower 95% confidence limits. For *Swietenia macrophylla*, percentage of surviving trees showing attack by the shoot borer *Hypsipyla grandella* is also indicated.

	Height	Diameter	Survival	Attacked
Swietenia with *I. edulis*	6.7 m (6.2–7.2 m)	5.1 cm (4.7–5.4 cm)	93%	33%
Swietenia with *I. ilta*	6.3 m (5.7–7.0 m)	5.8 cm (5.3–6.3 cm)	89%	33%
Swietenia alone (control)	4.2 m (3.8–4.8 m)	4.6 cm (4.1–5.1 cm)	84%	46%
Inga edulis	10.8 m (10.6–11.1 m)	13.4 cm (12.9–13.8 cm)	95%	–
Inga ilta	7.6 m (7.2–7.9 m)	10.6 cm (10.1–11.2 cm)	87%	–

In pairwise statistical comparison of treatment results, height of mahogany trees interplanted with both *Inga* species were significantly greater than for mahogany alone in the control plots (one-way ANOVA with Tukey pairwise comparisons of means; $p < 0.01$). Between the two *Inga* treatments, difference in height of mahogany was non-significant. For diameter growth, only for the treatment of mahogany grown with *I. ilta* was the mean diameter significantly greater than the control mahogany ($p < 0.01$), between the two *Inga* treatments, differences in mahogany diameter were also non-significant.

Survival of mahogany was greatest (93%) in the plots interplanted with *Inga edulis*, compared with the plots interplanted with *I. ilta* (89% survival) and with the control plots of mahogany alone (84%).

Incidence of attack of mahogany by the shoot borer *Hypsipyla grandella* was not observed for any of trees in the three treatments during the first 20 months following plantation establishment. The first evidence of shoot borer attack in the plantation occurred after maintenance cleaning of the plantation (cutting of competing weedy vegetation) was carried out at 20 months. After that time, incidence of shoot borer attack began to increase, and was most prevalent in the mahogany control plants. At 26 months following plantation establishment, 21% of the surviving mahogany plants in the control plots showed attack by the shoot borer, as compared with 13% of surviving

145

mahogany plants grown with *I. edulis* and 4% of those grown with *I. ilta* showing shoot borer attack. Attack of the mahogany plants by the shoot borer continued to increase for all treatments in subsequent months. At the end of the observation period, at 31 months following plantation establishment, attack by the shoot borer was observed in 33% of the surviving mahogany plants in both of the *Inga* treatments, while attack of the mahogany control plants was higher at 46% (Table 1).

DISCUSSION

The results clearly demonstrate a significant treatment effect of improved early growth of mahogany when interplanted with *Inga* species, in comparison with mahogany grown alone in the control plots. The growth rates for mahogany reported here compare very favourably with other trials of plantation-grown mahogany reported by Lamb (1966). The treatment effects are probably the result of a combination of several factors, including stimulation of upward growth of mahogany by the closer spacing of trees in the treatment plots, protection of the mahogany plants from the shoot borer by "hiding" them among the *Inga* plants, improvement of soil quality through the contribution of *Inga* leaf litter, and perhaps other factors. Pugnacious ants, which feed at foliar nectaries in *Inga* species, are known to provide protection to the host plants from herbivores (Bentley, 1977; Koptur, 1985). In the present case, it may be that the pugnacious ants feeding on *Inga* foliar nectaries also provide protection from herbivores to the adjacent mahogany trees, but we did not test for that hypothesis. We expect that soil improvement from the decomposition on *Inga* litter will be a more important factor at subsequent stages of growth of the plantation.

The significant differences in tree growth among the adjacent randomized experimental blocks in the former pasture (Table 2) appear to be due to microsite variation in soil quality, such as differences in severity of soil compaction in different areas of the pasture. In this respect, it is of interest to note that the growth rate of *Inga edulis* did not differ among blocks and apparently was unaffected by such differences in soil quality, whereas growth of *I. ilta* and mahogany was substantially reduced in Block 2 where the soil conditions appeared to be poorer.

Attack of the young mahogany plants by the shoot borer *Hypsipyla grandella* was negligible for all three of the treatments during the critical, early stages of growth, up to 20 months following plantation establishment. The first evidence of attack by the shoot borer occurred after the plantation was cleaned of competing woody vegetation, suggesting that the cleaning and disturbance caused the mahogany trees to be more exposed and "apparent" to shoot borers migrating from the surrounding forest region. In subsequent months, shoot borer attack increased for all treatments but was most severe for the mahogany trees grown alone in the control plots. It does appear, therefore, that the *Inga* plants did provide some protection from the shoot borer by making the mahogany plants less "apparent" to this specialized herbivore. Attack by the shoot borer, however, may not have been a significant factor in the differential mortality of mahogany plants in the control vs. treatment plots, because much of that mortality occured in the

TABLE 2. Variation in mean height and diameter at breast height of *Swietenia macrophylla, Inga edulis,* and *Inga ilta* at 2 years 7 months from plantation establishment, in the three trial blocks. Blocks 1 and 3 were located near the edge of the former pasture along fencerows where sparse woody vegetation existed amid the pasture grass, while Block 2 was located near the centre of the pasture with a vegetative cover almost entirely of grass. The site was grazed for about 20 years prior to establishment of the silvicultural trials, and the differences in growth rates of the trees may be related to increased soil compaction by cattle in the centre of the pasture. In contrast to the other two species, *Inga edulis* was unaffected by the presumed differences in soil quality among the trial blocks. In each of the three plots per block were 25 *Swietenia* trees, with 96 *Inga* trees in the treatment plots.

	Mean Height (m)			Mean Diameter (cm)		
	Blk 1	Blk 2	Blk 3	Blk 1	Blk 2	Blk 3
Swietenia with *I. edulis*	6.4	5.6	8.2	4.9	4.4	5.9
Swietenia with *I. ilta*	7.4	4.2	7.6	6.7	4.5	6.2
Swietenia alone (control)	5.9	2.9	3.2	5.9	3.5	3.6
Inga edulis	10.5	10.7	11.3	13.1	13.7	13.3
Inga ilta	8.7	5.3	8.7	11.6	8.8	12.7

first 20 months following establishment. The increase in shoot borer attack towards the end of the observation period certainly resulted in slower growth of the affected individuals, but it appears that trees over 5–6 m in height which are growing vigorously are likely to survive shoot borer attack at that stage of growth.

Further observations were made of the *Swietenia – Inga* silvicultural trial at 4 months after the quantitative results reported here (2 years 11 months from establishment of the plantation). At that stage, growth of mahogany in the plots with *Inga edulis* appeared to be slowing down considerably, because the crowns of the mahogany trees were shaded by *Inga* crowns. In the plots with *I. ilta*, in contrast, the crowns of the mahogany trees were more fully illuminated; the mahoganies exhibited continued vigorous growth and were beginning to grow up through and above the canopy of *Inga* (Fig. 2). This was enhanced by a silvicultural treatment carried out at 31 months and again at 35 months of plantation growth: using extendable pole pruners which can reach up to 10 m, the crowns of the *Inga ilta* trees were pruned where they contacted the crowns of mahogany, thus allowing for full illumination of the mahogany crowns. We believe that later growth of the plantation will demonstrate that *Inga ilta* is the preferred species to be interplanted with *Swietenia macrophylla,* because height growth is more nearly equal in the two

Fig. 2. Mahogany (*Swietenia macrophylla*) in experimental plot with *Inga ilta*, at 2 years 11 months after establishment of the plantation. The person in the photograph is holding the trunk of a mahogany, and the surrounding trees are all *Inga ilta*. The crowns of the *Inga* trees have been pruned with extendable pole pruners, to permit full illumination of the mahogany crown and allow the mahogany to grow above the *Inga* trees.

species, as compared with *Inga edulis* which grows much more quickly than *Swietenia* and will tend to shade it out. It is perhaps significant that although the mean height of mahogany grown with *I. edulis* is somewhat greater than for mahogany grown with *I. ilta,* the mean diameter is greatest for mahogany grown with *I. ilta.* This suggests that the mahogany trees growing in the shade may be in a somewhat "etiolated" condition, i.e. the shade-grown trees are relatively tall but thin — not a desirable condition for a timber production plantation where trunk diameter growth is very important for economic productivity of the stand.

Under plantation conditions, we have observed that *Inga ilta* trees begin to flower and fruit after about 4 years. The seeds of *I. ilta* are edible and rich in protein (see Pennington & Robinson, this volume), and the fruits are also edible as are those of all *Inga* species. A mixed plantation of *I. ilta* and mahogany, therefore, presumably could produce a cash crop of edible fruits and seeds for a number of years while the mahogany trees mature and attain a marketable size. We estimate that at least 30 years will be required for the mahogany trees to attain a marketable minimum diameter of 60 cm. At the spacing distance utilized in the present experiment, it may be necessary to thin out the stand eventually, removing most or all of the *Inga* trees and leave a pure stand of mahogany spaced at 8 m × 8 m.

The present and future high value of mahogany timber appears to justify efforts to grow this species under plantation conditions (Lamb, 1966). For local farmers in Amazonian Ecuador, the establishment of small woodlots of mahogany combined with *Inga* seems to be a viable economic activity, especially since the *Inga* trees can produce cash income while the mahogany matures. At the Amazon Plant Conservation Center of Jatun Sacha Biological Station, we are now working with local farmers to establish on-farm agroforestry trials, including *Swietenia – Inga* plots similar to the experiment reported here.

ACKNOWLEDGEMENTS

We thank the Liz Claiborne and Art Ortenberg Foundation for its long-term support of the Amazon Plant Conservation Center, and the Swedish International Development Agency and Barnens Regnskog (Children's Rainforest Sweden) for support during establishment of the silvicultural experiments. Terry Pennington helped with the experimental design and provided welcome advice and encouragement. The Jatun Sacha staff and numerous volunteers assisted in the plantation establishment and data collection.

REFERENCES

Bentley, B.L. 1977. Extrafloral nectaries and protection by pugnacious bodyguards. Annual Rev. Ecol. Syst. 8: 407–427.
Gullison, R.E., Panfil, S.N., Strouse, J.J. &. Hubbell, S.P. 1996. Ecology and management of mahogany (*Swietenia macrophylla* King) in the Chimanes Forest, Beni, Bolivia. Bot. J. Linn. Soc. 122: 9–34.

Holdridge, L.R. 1967. Life Zone Ecology. Tropical Science Center, San José, Costa Rica.

Koptur, S. 1985. Alternative defense against herbivores in *Inga* (Fabaceae: Mimosoideae) over an elevational gradient. Ecology 66: 1639–1650.

Lamb, F.B. 1966. Mahogany of Tropical America: Its Ecology and Management. University of Michigan Press, Ann Arbor, USA.

Pennington, T.D. 1997. The Genus *Inga*: Botany. Royal Botanic Gardens, Kew, England.

CHAPTER 10. UTILIZATION PROFILE OF A NEW SPECIES: *INGA ILTA* T.D. PENN.

T.D. PENNINGTON & R.K. ROBINSON

The centre of diversity of *Inga* is in the Andean foothills of Amazonian Colombia, Ecuador and Peru, with over 100 species already described and an estimated further 30 undescribed species at present known only from incomplete data (Pennington, 1997). The far west of Amazonia, which includes parts of Colombia, Ecuador, Peru and Brazil, is already known as a centre of domestication of food plants (Clement, 1989). Examples include *Pouteria caimito* (*Sapotaceae*) and *Quararibea cordata* (*Bombacaceae*).

The best known species of *Inga* which is cultivated for its fruit is *I. edulis*, and it is possible that this species also originated in the same area, as some of the best forms known today are found in the Peruvian Amazon. For example, in the neighbourhood of Yurimaguas, Department of Loreto, Peru, some forms of the fruit are up to 2 m long (Erick Fernandes, pers. comm.).

The discovery of *I. ilta*, cultivated by the Quichua Indians in the provinces of Napo and Pastaza in the Ecuadorean Amazon, provides another example of domestication from the same region. That this large-fruited species has been overlooked for so long is probably due to the confused state of *Inga* taxonomy, where until recently all *Inga* trees grown in gardens were regarded as *I. edulis*.

RELATIONSHIPS OF *I. ILTA*

Inga ilta was first recognized as a new species in 1991 when fruiting specimens were collected in the vicinity of Jatun Sacha Biological Station, Napo Province, Ecuador. Later searches of the major herbaria during a monographic study of the genus revealed a few earlier flowering collections from other localities nearby.

Inga ilta is so far known only from limited areas of Napo Province in Amazonian Ecuador, and from Iquitos in Amazonian Peru, where it occurs below 200 m altitude. Nearly all the collections are from cultivated or protected trees in house gardens. The few collections from wild populations are from Amazonian Peru, in a region with an annual precipitation around 3000–4000 mm, and mean annual temperature 26°C. In the areas where it is cultivated in Ecuador the annual precipitation may reach 5000 mm, and the rainfall is distributed throughout the year.

The leaf and floral morphology of *I. ilta* is similar to that of the widespread *I. punctata* (sect. *Pseudinga*), which occurs in the same area, but *I. ilta* is distinctive because of its silky indumentum on the lower surface of the leaflets, the unusual sunken foliar nectaries, and because of the massive legume up to 100 cm long, 5–6 cm wide and 3–4 cm thick (Figs. 1 & 2). It is fairly certain that *I. ilta* is most closely related to *I. punctata*, but whether it is derived from it cannot be inferred from a study of their external morphology alone. An investigation at the level of isoenzymes or molecular markers might be more revealing.

FIG. 1. *Inga ilta* (scale 16 cm) (photo. T.D. Pennington).

Inga punctata occurs throughout Central America, Andean South America and across Amazonia. It has a wide altitudinal range (sea level to 2000 m) and a wide range of ecological tolerance. It is usually found in disturbed vegetation along roadsides, in pasture and on riverbanks. It also has a wide tolerance of climate, ranging from the everwet Amazonian slopes of the Andes with 5000 mm or more of annual rainfall to the strongly seasonal Pacific slopes of Central America with less than 2000 mm precipitation. It is a light-demanding gap colonizer of rain forest, on both non-flooded and seasonally flooded land. *Inga punctata* is widely used as a shade tree for coffee throughout its range.

152

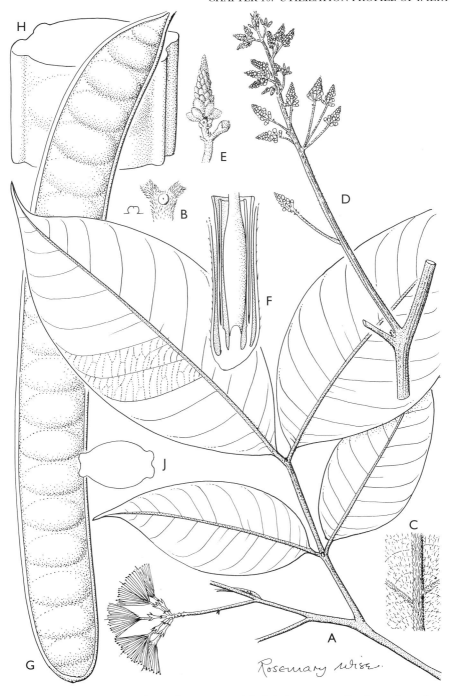

FIG. 2. *Inga ilta*. A, habit × ²/₃; B, foliar nectary × 2.6; C, indumentum of leaflet undersurface × 3; D, inflorescence × ²/₃; E, enlargement of flower buds × 1.3 (*Pennington & Freire* 13772); F, flower section × 13.3 (*Madison et al.* 5468); G, legume × 0.3; H, legume section × ²/₃; J, legume section × 0.3 (*Pennington & Freire* 13772).

153

A common pattern of distribution among Amazonian trees is for a widely distributed species to have closely related but local 'satellite' species along the western rim of Amazonia. The widely distributed *I. punctata* conforms to this pattern with several local satellites in west and southwest Amazonia. These are *I. ilta* in Ecuador and northern Peru, *I. longifoliola* in Colombia and adjacent Brazil and *I. peduncularis* in NE Bolivia and adjacent Brazil. This pattern is repeated across unrelated families such as *Meliaceae*, where there are examples in *Trichilia* (Pennington & Styles, 1981: 20) and in the *Sapotaceae* (Pennington, 1990: 39). Examination of within-species variation of widely distributed species in tropical America also confirms that western Amazonia is a centre of present day speciation.

UTILIZATION

Inga ilta was included in the ODA agroforestry trials at Jatun Sacha in Amazonian Ecuador (1991–1996) and it was also used at a site 2 km from Jatun Sacha in a mixed trial with *Swietenia macrophylla* (Broad-leaf Mahogany), and it soon became apparent that it had several potentially useful growth characteristics. Full details of its growth and biomass production are given in Pennington, and Murphy & Yau, this volume, but its areas of possible use are summarized here.

Growth and competitive ability-shading
Inga ilta is one of the fastest growing Amazonian *Inga* species yet encountered, exceeded only by the well known *I. edulis*. The growth figures from the ODA trial at 3 m × 3 m spacing (at 2 years 6 months after establishment) were mean height 9 m, mean diameter 12.1 cm. At a nearby site when grown in old compacted pasture at 4 m × 4 m spacing (aged 2 years 7 months) the mean height was 7.6 m, and mean diameter 10.6 cm.

Although not quite as vigorous as *I. edulis*, in agroforestry terms it has significant advantages in its habit. It is more profusely branched than *I. edulis*, and its leaflets are larger and much thicker. It therefore casts a deeper shade and the leaves, which take longer to break down, rapidly form a long-lasting and thick leaf litter. In the shading experiments at Jatun Sacha (Pennington, this volume), *I. ilta* was the best of all species in the trial, giving total elimination of all secondary vegetation including *Brachiaria, Vernonia, Cecropia* and other fast growing woody and herbaceous species, within 18 months of establishment. In the mixed trial with *Swietenia* (Neill & Revelo, this volume) in old compacted pasture, it soon dominated the vigorous *Brachiaria decumbens*, a species introduced from Africa for cattle fodder and now covering huge areas of abandoned land in lowland Amazonia. Other species, such as *I. edulis* with greater height and diameter growth are less effective in this respect because of their more open branching habit and thinner, lighter and smaller leaflets. *Inga ilta* would therefore seem to be a good candidate for bringing back this type of abandoned site into productive use.

Nurse Crop for Timber Species

Neill & Revelo (this volume) describe the use of *Inga* as a nurse crop for mahogany (*Swietenia macrophylla*) — it was found to significantly enhance the growth of that species compared with the control. The reasons for this improvement are uncertain, but a combination of several factors would seem to account for it:

(1) The rapid formation of a layer of organic material from the decomposing leaf litter which enables rooting into the organic material above the compacted mineral soil to a condition resembling the floor of natural forest.
(2) Stimulation of vertical growth by placing between *Inga* trees. *Inga ilta* seems ideal for *Swietenia* because their growth rates are similar and unlike the more vigorous *I. edulis*, the light demanding *Swietenia* is never overtopped and shaded.
(3) Reduction of soil temperature below the leaf litter layer to levels near those of a forest soil.
(4) Biological control of the *Swietenia* shoot borer by insects attracted to the *Inga* foliar nectaries. It is probably significant that the lowest level of *Hypsipyla* infection (Neill & Revelo, this volume) was on the *Swietenia* grown with *I. edulis*, as this species has much larger and more active nectaries than does *I. ilta*, and is known to attract a wide range of insects including ants and wasps (see Ackerman *et al.*, this volume).

Alleycropping

The ability of *I. ilta* to establish and flourish on poor acidic compacted soils has been shown above. For alleycropping the additional requirements are the capacity to withstand repeated coppicing and production of large amounts of branch and leaf biomass under a coppicing regime. In the ODA trials at Jatun Sacha, *I. ilta* was coppiced at 6 monthly intervals, without any mortality, so long as one or two leafy branches were left unpruned at each coppicing. Its production of branch and leaf biomass was 13.8 tonnes/ha/year (dry weight) at a spacing of 3 m × 3 m. This is more than any other species in the trial, including *I. edulis*. Although not grown in this instance in an alleycropping configuration, there is no reason to believe that it would perform less well at higher density planting, given the experience with other species, such as *I. edulis* and *I. oerstediana*.

Fuelwood

The productivity of each species in the Jatun Sacha trial was calculated at the end of 3 years after establishment (Pennington, this volume) when a sample of trees from each plot was felled and weighed to calculate the production of wood biomass. The range of wood biomass encountered for all *Inga* species was from 9 to 25 tonnes/ha/year, and *I. ilta* with 18.6 tonnes had the greatest production apart from *I. edulis* (25 tonnes). *Inga ilta* is in the same range as *I. edulis* for calorific value, density and ash content (Murphy & Yau, this volume). As the latter species is one of the most popular providers of fuelwood for cooking, it would seem therefore that *I. ilta*, would be a useful woodlot species for use on small areas of marginal land, which might not be suitable for cultivation of any other crops.

Food

Inga ilta flowers 4 years after establishment (D. Neill, pers. comm.), but it is not yet known whether it produces a regular fruit crop. Other species in the same geographical area, such as *I. edulis* and *I. densiflora*, flower and fruit once or sometimes twice a year. The cultivated form of *I. ilta*, which is commonly found in gardens of the local Quichua people, has a very large legume containing up to 25 seeds, each surrounded by a copious pulp. The large size of the seeds (c. 4 × 2.5 cm) ensures that one legume provides a very large 'snack', more than enough for one person.

The edible sarcotesta has the consistency of candyfloss and at maturity is easily detached from the embryo. It is composed of a slender network of cellulose fibres and water (81%), but its attractive sensory properties arise from the presence of sugars (15.4%). A small amount of protein (1.2% calculated as crude protein) may be present as well. As the total intake of this material per adult/child per day will be limited, any impact on the diet is likely to be minimal. Nevertheless, the role of any foodstuff which encourages social activity within a community or family should not be ignored.

The actual embryos are cooked by placing them in a pan, covering with water and bringing them to the boil. The embryos are gently simmered for 40–45 minutes before the water is drained off and discarded. These cooking practices are probably essential to improve the nutritional value of the embryos by removing any toxic principles that may be present, e.g. trypsin inhibitors or haemagglutinins. In any event, the embryos would be unpalatable if consumed raw, whereas the cooked material is readily eaten hot or cold at main meals, usually in soups.

The chemical composition of a typical sample of cooked embryos is shown in Tables 1 and 2 and the comparison with the broad bean (*Vicia faba*) from Europe shows some interesting differences.

In this context, 'cooking' was taken to mean boiling in unsalted water for 10 minutes as specified for broad beans and no attempt was made to ascertain the palatibility of the embryos. Consequently, it should be noted that the figure for crude protein (N × 6.25) might be reduced a little by more extensive cooking and some minerals and water-soluble sugars might also leach out. Nevertheless, the food value of the embryos is evident in that both the calculated calorific value (165.4 kcal/100 g) and protein level are well above the figures for broad beans. Some important minerals, such as calcium (110 mg/100 g), magnesium (50 mg/100 g) and iron (33 mg/100 g), are also above the levels quoted for broad beans, namely 56, 36 and 16 mg/100 g respectively.

It is likely that the embryos of *Inga ilta* would, with the possible exception of carotene, provide little in the way of vitamins, but this conclusion is typical of most legumes.

Non-replicated samples of *I. ilta* embryos gave a mean fresh weight of 11 g, and a mean number of 22 seeds in each legume. A well-grown 5–6 year old tree could be expected to bear at least 100 fruits, which would provide 24 tonnes per ha assuming a final spacing of 10 m × 10 m.

Overall, it would seem that this crop can provide a welcome input to the diet of those with access to the maturing fruits. Whether excess fruits could be dried for consumption and/or trading out of season deserves attention, particularly if the acreage devoted to the crop is to be expanded.

TABLE 1. Analysis of the cooked embryos of *Inga ilta*, together with published data for *Vicia faba** cooked under similar conditions; all figures as g/100 g of crop as consumed.

Crop	Water	Protein	Fat	Crude fibre	Starch	Water-soluble Carbohydrates
Inga (cooked)	57.7	13.5	0.2	1.2	23.2	4.2
Vicia (cooked)	73.7	7.9	0.6	6.5	10	1.3

(* after Holland *et al.*, 1991)

TABLE 2. Analysis of the cooked embryos of *Inga ilta*, together with published data for *Vicia faba** cooked under similar conditions.

	Calorific value	Calcium	Mg	Fe
Inga	165.4 kcal/100 g	110 mg/100 g	50 mg/100 g	33 mg/100 g
Vicia	48 kcal/100 g	56 mg/100 g	36 mg/100 g	16 mg/100 g

(*after Holland *et al.*, 1991)

ACKNOWLEDGEMENTS

This study was financed by the British Government Overseas Development Administration, Projects R.4729 and R.6075 of their Forestry Research Programme.

REFERENCES

Clement, C.R. 1989. A Center of Crop Genetic Diversity in Western Amazonia. Bioscience 39(9): 624–631.

Holland, B., Welch, A.A., Unwin, I.D., Buss, D.H., Paul, A.A. & Southgate, D.A.T. 1991. McCance & Widdowson's "The Composition of Foods." (Ed. 5). Royal Society of Chemistry and Ministry of Agriculture, Fisheries and Food, London.

Pennington, T.D. & Styles, B.T. 1981. *Meliaceae*. Fl. Neotrop. Monogr. 28.

Pennington, T.D. 1990. *Sapotaceae*. Fl. Neotrop. Monogr. 51

Pennington, T.D. 1997. The Genus *Inga*, Botany. Royal Botanic Gardens, Kew.

CHAPTER 11. *INGA* MANAGEMENT

T.D. PENNINGTON

Inga species are in many ways the ideal subject for low cost nursery cultivation in the humid tropics, easy to germinate, with rapid growth and relatively free from pests and diseases. This chapter describes the management requirements based on our experience of the recent *Inga* trials which have been established in Honduras, Costa Rica, Ecuador and Peru.

SEED COLLECTION

Seed collection methods for trials, plantation work or for agroforestry are generally designed to encompass as much genetic variation as possible. A minimum of 30–50 parent trees from each provenance has been recommended with a wide parent tree spacing of at least 100 metres (Hughes, 1987). However, in the case of *Inga* there are several practical problems which prevent this method being followed, such as the difficulty of finding sufficient trees of a species in one locality. Most rain forest species of *Inga* occur at low density (often around 1–2 individuals per hectare) and are therefore difficult to find. They also produce irregular seed crops which may not be fully synchronised, and a good seed year may be followed by several poor seed years. The period of fruit maturity is very short (days rather than weeks), and the seeds are subject to heavy predation from birds and from the attention of dispersing animals (primates). Fruit knocked to the ground by dispersers and predators is quickly removed by other animals, especially rodents. Also the seed of some species is subject to predation by bruchid beetles. In practice, therefore, it is often difficult to obtain seed from more than a few parent trees, which may yield only a few hundred seeds.

The fleshy *Inga* seeds are surrounded at maturity by a sweet edible pulp, which rapidly ferments on exposure to air. Removal of the pulp reveals the naked embryo, which, due to its soft and fleshy texture, is liable to dry out rapidly, causing loss of viability.

When collecting seed, only those which are clean, healthy and free from insect damage should be chosen. Because of the attraction of *Inga* fruit for birds and primates, it is best to collect the seed when still slightly immature, before it attracts the attention of animals. Collecting strategy should be planned well in advance, and the trees kept under observation for several weeks before collection, as a single week can see a large tree bearing many hundreds of fruit stripped bare. On the eastern slopes of the Andes, where there is little or no dry season, the peak fruiting season is from November to March; while in seasonal forests, as in Pacific Central America, fruiting usually occurs at the end of the dry season or beginning of the wet season.

Seed Storage

Like the fleshy seeds of many rain forest trees, those of *Inga* are intolerant of dessication (so-called recalcitrant seeds) and therefore have a very short period of viability under natural conditions. In large-seeded species such as *I. sapindoides* or *I. oerstediana*, germination often starts within the unopened legume, by which time the fleshy pulp surrounding the seed has begun to ferment. At ambient temperature the seeds will retain viability for a maximum of one or two weeks within the unopened legume. However, it is possible to extend this period to around a month by lowering the temperature at which the fruit is stored (minimum 10–15°C).

Once removed from the fruit, the seeds with their fleshy pulp will quickly deteriorate and rot. If storage for more than a few days is required, the pulp must be removed from the seed as soon as possible, certainly before it begins to ferment. Otherwise, the naked embryo will be infected by fungal spores and will not be amenable to surface treatment with a fungicide. The embryo is highly susceptible to dessication and will die if the moisture content is allowed to fall below 40–45% (Pritchard *et al.*, 1995), so they should be kept in closed containers made of plastic or glass, but with the stopper loose in order to allow some gaseous exchange. Embryos removed from clean fruit in which fermentation has not started should be washed in clean water and then allowed to stand in a shady well-ventilated place until the excess water has evaporated.

Efforts to store the fresh seeds of *I. edulis* and *I. striata* at ambient temperature were reported by Castro & Krug (1950) and Castro (1952). These authors treated the seed with a dilute solution of potassium hydroxide to remove the sarcotesta, followed by immersion in a dilute solution of glycerine. They reported good viability after several months storage in glass containers. Zúniga (1996) treated *Inga* embryos with a dilute solution (125 ppm) of a proprietary fungicide (thiabendazole) for 15 minutes, after which they were allowed to dry to remove surplus liquid. Embryos treated in this way were kept in sealed glass containers for 80 days at ambient temperature, and retained a high percentage germination during this period. Further details of this storage technique, which was used with *I. leiocalycina*, can be found in Zúniga (1996).

Propagation

1) *Seed*

When propagating *Inga* by seed it is first necessary to establish a small nursery to provide the developing seedlings with the most favourable conditions for survival and growth. Germination and subsequent growth require shade and a source of water, so the location of the nursery must take this into consideration. Shade can be provided by trees, and this is the preferred method as it requires no labour, but has the disadvantage of falling leaves which may have to be removed frequently from the seed bed. The alternative is to cover the seed beds with a wooden framework which is thatched with palm leaves. This gives the seedlings better protection from heavy rain but needs regular maintenance if it is not to deteriorate rapidly.

The whole nursery area may have to be fenced to protect the plants from damage by animals (e.g. goats, cattle, rabbits).

Preparation of seed before sowing

The fleshy pulp must be removed before sowing and this is facilitated by immersion in water. Once the pulp is removed the naked seeds are then immersed for 12 hours in water containing macerated root nodules taken from mature *Inga* tree roots, together with some forest soil taken from beneath an *Inga* tree. The purpose of this treatment is to infect the seeds with nitrogen fixing bacteria and with mycorrhizal fungi (Fernandes, this volume). So far, all species trialled have been found to nodulate.

Inga seeds vary in size from about 0.5–1 cm long (*I. marginata*) to 5 cm long (*I. edulis*). The largest seeds like those of *I. edulis* should be planted about 1–2 cm below the soil surface whilst the smallest seeds should be covered only by a fine layer of soil.

The preferred method of propagation is by sowing individual seeds into polythene bags filled with forest soil. This method ensures a high percentage germination, rapid subsequent growth and very little disturbance of the roots when the seedlings are transplanted to open ground. The black polythene bags need to be at least 15 cm diameter and 20 cm deep, to accommodate the young plants for 3–4 months.

If sufficient seed is available, two can be planted in each polythene bag and at a later date the more vigorous seedling can be selected and the weaker one discarded. Local forest soil high in organic matter should be used in the polythene bags. In moist shady conditions a high percentage (95–100%) germination will occur in one to two weeks, seedlings should be ready for transplanting when 30–40 cm high after three to four months. During this period shade is gradually removed until the seedlings can tolerate full sun.

The seed beds on which the polythene bags are placed should measure about 10 m long and 1 metre wide and are slightly raised (5–6 cm) above the surrounding soil level to avoid waterlogging or flooding in heavy rain. The surface of the seed bed is covered with a sheet of black polythene which will prevent the seedlings from rooting through the base of the polythene bag into the seed bed. Plants which have rooted into the seed bed will be severely checked if the roots are broken when the bags are moved for transplanting.

If polythene bags are unobtainable or cannot be afforded, then seed can be successfully planted directly into a seed bed. The resulting bare root *Inga* seedlings can be transplanted without adverse effects, so long as the root systems are kept moist in a bucket of water after the seedlings are dug up from the beds.

2) *Root suckers*

A few species of *Inga*, notably *I. vera* and *I. sapindoides* are known to produce sucker shoots from surface roots. These are easily removed with a small section of root from the parent tree. Under normal circumstances only a few suckers are found on each tree and further work needs to be done to investigate whether sucker production on roots can be promoted by artificially damaging the surface roots.

3) *Stem cuttings*

Experimental work over the past few years has shown that *Inga* species can be easily propagated from semi-ripe branch cuttings. Among species which have been successfully propagated by this method are *I. calderonii, I. edulis, I. laurina, I. marginata, I. oerstediana, I. punctata, I. sapindoides* and *I. subnuda.* It seems that the only species which are difficult to propagate in this way are those with very slender branches, such as *I. heterophylla.* The cuttings should be approximately 15 cm long and up to 1 cm diameter with two leaves. The lower leaf is removed cleanly and the stem cut cleanly about 1 cm below the insertion of the lower leaf node. The upper leaf should be cut back to 1 pair of leaflets to reduce water loss, or the leaf can be removed entirely. The base of the cuttings can be treated with a proprietary hormone rooting powder such as IBA which contains a fungicide, though rooting can be achieved without this. It appears that the fungicide is probably more important than the rooting hormone. The cuttings are then plunged to halfway in a 50/50 mix of coarse sand and organic soil or in 100% coarse sand. After thorough watering the cuttings are covered with a thin film of transparent polythene to conserve moisture and then placed in a shady place at ambient temperature. The containers used to hold the cuttings should be free draining. Most species root easily within 3–4 weeks. As soon as rooting is observed the cuttings should be carefully transplanted into plastic bags measuring at least 15 cm diameter and 20–25 cm deep.

This method has been used under glasshouse conditions at Kew, England and at Cornell University, U.S.A. and also under field conditions in Honduras (Zúniga, pers. comm.). *Inga punctata* rooted readily in forest soil in Honduras, without the use of hormone rooting powder, as long as the cuttings were protected by a polythene sheet to conserve moisture during the rooting period. *Inga* plants are very vigorous and it is important to use a large enough container to allow adequate development of the strong root system. Rooted cuttings should be placed under shade for 2–3 weeks after which they can be gradually introduced to full sunlight.

Studies of wild populations of *Inga* have shown that species are self incompatible, and that fruit set increases with increasing distance between the pollinating trees (Koptur, 1983, 1984). It is important therefore to use cuttings from several, preferably widely separated, individuals, if good fruit set is a requirement.

Transplanting

Seedlings are generally transplanted at 3–4 months old when they are 30–40 cm high. The holes to take the *Inga* plants should all be dug before the transplanting commences so that the seedlings can be planted with the minimum delay and without disturbance to the roots. Ideally this should be done after rain and on a cloudy day. If the plantation is in an area with a seasonal climate then transplanting should be carried out at the beginning of the rainy season. Seedlings planted in the dry season are liable to be seriously retarded. For the first year the site will need to be weeded 3–4 times to eliminate competition which would swamp the *Inga* plants. If the competing vegetation is sparse then it is generally sufficient to clean an area of about 1 square metre around each *Inga* seedling.

Weeding must continue until such time as the *Inga* plants are forming a crown which begins to suppress competing vegetation, usually between 1 and 2 years.

The spacing for transplanting seedlings depends on the purpose of the cultivation. Different species are suited to different systems and will require different spacing. For alley cropping the usual design is 4 metres between rows and 50 cm between plants. For the recuperation of old pasture a vigorous species, such as *I. oerstediana*, will require a spacing of 4 m × 4 m in alternating rows. Less vigorous species, such as *I. marginata*, would be planted 3 m × 3 m apart. *Inga* planted on marginal land for fuelwood production should not be planted closer than 5 m × 5 m.

When used for shade over coffee, the *Inga* plants are generally spaced at 8 m × 8 m or 10 m × 10 m, and they are generally planted at the same time as the young coffee plants.

SELECTION OF SPECIES

With such a wide range of species available, they must be carefully chosen to suit the local ecological and climatic conditions and their required usage. Widespread and ecologically variable species, such as *I. oerstediana*, have provenances differing widely in their characteristics and requirements. Observations on the behaviour of *Inga* species in wild populations and under trial conditions lead to the conclusion that usually the best strategy is to collect seed from local populations which are known to be adapted to local conditions. There is a high risk involved in introducing species or provenances from geographically distant areas. Studies on the ecological requirements of wild species and of the performance of *Inga* species in trials have provided a lot of information on species requirements. The species listing that follows is according to ecological preferences and uses, and is restricted mostly to those with some actual or potential use in agroforestry or forestry systems.

1) *Ecological Adaptations*

1. Species adapted to non-flooded, poor acid soils in the lowlands.

   a) Central America: *I. densiflora, I. edulis, I. marginata, I. oerstediana, I. samanensis, I. thibaudiana*.
   b) South America: *I. cayennensis, I. densiflora, I. edulis, I. heterophylla, I. marginata, I. oerstediana, I. samanensis, I. thibaudiana, I. umbratica, I. velutina, I. vismiifolia*.

2. Species adapted to white sands in Amazonia: *I. ingoides, I. heterophylla, I. pilosula*.

3. Species adapted to periodically flooded or poorly drained sites.

   a) Central America: *I. coruscans, I. vera*.
   b) South America: *I. lopadadenia, I. duckei, I. cinnamomea, I. nobilis* ssp *nobilis, I. stenoptera, I. cecropietorum, I. velutina, I. vera, I. ingoides, I. macrophylla*.

163

4. Species of montane forest.

   a) Central America: *I. ruiziana* (to 2000 m), *I. densiflora* (to 1500 m), *I. punctata* (to 1500 m), *I. oerstediana* (to 1500 m).
   b) South America: *I. tomentosa* (to 1500 m), *I. marginata* (to 2000 m), *I. ruiziana* (to 2000 m), *I. densiflora* (to 2000 m), *I. punctata* (to 2000 m), *I. setosa* (to 1500 m), *I. fendleriana* (to 2400 m), *I. ornata* (to 2200 m), *I. adenophylla* (to 2200 m), *I. oerstediana* (to 1700 m), *I. insignis* (to 3000 m), *I. striata* (to 2000 m), *I. velutina* (to 1600 m), *I. balsapambensis* (to 1800 m), *I. villosissima* (to 2300 m).

5. *Inga* species adapted to a strongly seasonal climate with a rainless season of several months. Although all *Inga* species are evergreen, a few are found in semi arid conditions as for example on the Pacific coast of Central America and Ecuador. In these conditions they can be important shade species.

   a) Central America: *I. laurina, I. punctata, I. vera, I. oerstediana.*
   b) South America: *I. punctata, I. manabiensis, I. vera, I. oerstediana, I. adenophylla, I. carinata, I. insignis, I. feuillei, I. jaunechensis.*

2) *Uses*

1. Fruit trees. All *Inga* species have an edible sarcotesta surrounding the seed, but in the majority it is thin with little to eat. The species detailed here are the best eaters with large fruit containing a lot of flesh. Two species (*I. jinicuil* and *I. ilta*) have large edible embryos which are cooked and eaten in soups (see Pennington & Robinson, this volume).

   a) Mexico and northern Central America: *I. jinicuil.*
   b) Southern Central America (Costa Rica, Panama): *I. densiflora, I. spectabilis, I. jinicuil, I. edulis* (introduced).
   c) Pacific South America (including high Andes): *I. densiflora, I. spectabilis, I. ornata, I. edulis* (introduced), *I. insignis, I. feuillei.*
   d) Amazonian South America: *I. cinnamomea, I. densiflora, I. ilta, I. ingoides, I. edulis, I. macrophylla.*
   e) South east Brazil: *I. edulis, I. striata.*

2. Shade trees. The use of *Inga* for shade extends throughout tropical America from Mexico to Argentina. The greatest area of *Inga* is over coffee, but it is also widely used for cacao and on a smaller scale for tea. Although on large coffee growing concerns the tendency nowadays is to plant coffee without shade (see León, this volume) elsewhere on low cost enterprises, as in much of Honduras, it is still very popular for this use and is highly valued, not just for shade, but for its many ecological properties (Lawrence & Zúniga, 1996). The following listings are restricted to those species in each area which are the preferred ones for shade and are deliberately cultivated for that purpose. Many other species of *Inga* may be found as odd individuals in plantations, but these are mostly remnants of the original vegetation cover,

recognized by the farmer as *Inga* and managed for their shade. Most coffee growers propagate their own *Inga* plants, using seed from their own plantations and from surrounding areas.

a) Mexico and Central America: *I. cocleensis, I. densiflora, I. calderonii, I. edulis, I. oerstediana, I. jinicuil, I. punctata, I. vera, I. sapindoides.*

b) Andean South America: *I. carinata, I. leiocalycina, I. punctata, I. oerstediana, I. adenophylla, I. edulis, I. densiflora, I. ursi, I. setosa.*

c) West Indies. *I. laurina, I. vera.*

d) South East Brazil: *I. edulis, I. striata, I. vera.*

3. Fuelwood. The fuelwood characteristics of some *Inga* species are discussed by Murphy & Yau, this volume. The main use of the fuelwood is for domestic cooking, but it is also reported to be used for charcoal production in some countries (e.g. Honduras, Ecuador and Trinidad), though no data is available on the latter. *Inga* fuelwood also has a small scale industrial use in Honduras, where the prunings from the shade trees are used to fire the furnaces drying the coffee.

a) Central America: *I. marginata, I. ruiziana, I. densiflora, I. punctata, I. vera, I. oerstediana.*

b) South America: *I. densiflora, I. edulis, I. oerstediana, I. punctata, I. ilta, I. capitata, I. vera, I. adenophylla, I. ingoides.*

4. Eradication of old pasture and weed control. (see Pennington, and Hands, this volume). Species for this use have to be vigorous growers with a dense crown of thick leaves. Our knowledge of the use of *Inga* in this area is at present restricted to the few species which have been trialled or used in alley cropping experiments. The list that follows will no doubt grow as further species are investigated. The most promising plants that should be investigated are the species of section *Inga* (Pennington, 1997), in particular a group of ten or fifteen species related to *I. edulis* and *I. oerstediana*. Species in this section are some of the most vigorous and competitive and are well adapted to a range of ecological conditions and a wide altitudinal range.

a) Central America: *I. marginata*, profusely branched habit and vigorous growth on poor acid soils; *I. oerstediana*, very vigorous with wide altitudinal range; *I. punctata*, on richer soils; *I. edulis*, very vigorous, on poor soils.

b) South America: *I. marginata*, good over an altitudinal range of 0–2000 m; *I. oerstediana*, adapted to a wide range of ecological conditions, and wide altitudinal range 0–2000 m; *I. edulis*, best below 1000 m altitude on poor acid soils; *I. ingoides*, flourishes on white sands; *I. nobilis* and *I. vismiifolia*, good in compacted pasture, but may be too slow growing; *I. ilta*, very vigorous, profusely branching, with large leathery leaflets which are slow to decompose, very good in compacted pasture.

5. Species for alley cropping. These need to be tolerant of repeated coppicing at a height of 1–1.5 m above ground, and also profusely branched and capable of producing enough foliage to provide a permanent leaf mulch, for nutrients and weed control (see Hands, this volume).

  a) Central America. *I. marginata, I. oerstediana, I. edulis, I. samanensis* (acid soils); *I. punctata,* (richer soils).
  b) South America: *I. marginata, I. ilta, I. edulis, I. oerstediana, I. ingoides* (on white sands).

6. Honey producers. Several species (especially *I. marginata*) are already used by beekeepers, but there are many other species which produce a mass flowering over a period of several weeks. They are mostly found in sections *Pseudinga* and *Bourgonia* (Pennington, 1997) and are characterized by large spicate inflorescences or compound inflorescences bearing numerous small scented flowers.

  a) Central America: *I. coruscans, I. laurina, I. marginata, I. nobilis.*
  b) South America: *I. tomentosa, I. lopadadenia, I. marginata, I. cinnamomea, I. vismiifolia, I. nobilis.*

7. Species for forage. No analyses have been done to study the value of *Inga* species as providers of cattle fodder, but personal observations of the author and anecdotal evidence from various sources indicates that they have some potential. The only records are from South America: *I. heterophylla* (Bolivia), *I. macrophylla* (Peru, Bolivia).

8. Ornamental Species. Some species of smaller stature, and with plentiful flowering over a long period, would make useful ornamentals for parks and gardens. Some, such as *I. nobilis*, and *I. marginata*, are already used for this purpose. *Inga insignis* is cultivated as a street tree throughout Quito, Ecuador, where it is clearly tolerant of severe atmospheric pollution. Other species which should be tried as ornamentals are listed below.

  a) Central America: *I. laurina.*
  b) South America: *I. umbellifera, I. tenuistipula, I. cinnamomea, I. vismiifolia, I. pilosula, I. setosa, I. pluricarpellata, I. steinbachii.*

REFERENCES

Castro, Y.G.P. de & Krug, H.P. 1950. Experiências sôbre germinacâo de sementes de *Inga edulis*, espécie usada em sombreamento de cafeiros. Mimeograph, Forestry Service, São Paulo.

Castro, Y.G.P. de. 1952. Experiências sôbre germinaçao de sementes de *Inga striata*. Mimeograph, Forestry Service, São Paulo.

Hughes, C. 1987. Intensive study of multipurpose tree genetic resources. ODA research scheme R.4091, Final Report.

Koptur, S. 1983. Flowering phenology and floral ecology of *Inga* (*Fabaceae*: *Mimosoideae*). Syst. Bot. 8(4): 354–368.

Koptur, S. 1984. Outcrossing and pollinator limitation of fruit set: breeding systems of neotropical *Inga* trees (*Fabaceae*: *Mimosoideae*). Evolution 38(5): 1130–1143.

Lawrence, A. & Zúniga, R. 1996. The role of farmers' knowledge in agroforestry development: a case study from Honduras and El Salvador. Agricultural Extension and Rural Development Department (AERDD), Working Paper 96/5, The University of Reading.

Pennington, T.D. 1997. The Genus *Inga* – Botany. Royal Botanic Gardens, Kew.

Pritchard, H.W., Haye, A.J., Wright, W.J. & Steadman, K.J. 1995. A comparative study of seed viability in *Inga* species: desiccation tolerance in relation to the physical characteristics and chemical composition of the embryo. Seed Sci. & Technol. 23: 85–100.

Zúniga, R. 1996. Chemical treatment to improve the viability of recalcitrant *Inga* seeds. Unpublished M.Sc. thesis (Agroforestry), Cranfield University, Silsoe College, U.K.